A CONCEPTION OF THE UNIVERSE
— OR —
EXPLORING
THE UNIVERSE FROM WITHIN

Written By
JEFFREY C. VOGT

Ψ

PRODUCTION

A CONCEPTION OF THE UNIVERSE
or
EXPLORING THE UNIVERSE FROM WITHIN

By: Jeffrey C. Vogt

Copyright © 2015

ASIN: B012MJSVIE

ISBN-13: 978-1522781394

ISBN-10: 1522781390

All rights reserved. Designed in the United States of America. No part of this book, in any format, may be used or reproduced in any manner in print, audio, or electronic form without written permission from copyright holder except in the case of brief quotations embodied in critical articles or reviews.

conception.universe@gmail.com

Book and Cover design by F.N.G.
Cover photo credit: J.C.V. Transit of Venus June 5, 2012©
First Edition: August 21, 2015

To my wonderful wife and daughters.

"Daddy, why is the moon following us?"

An overelaborated, armchair perspective of a possible Universe.

CONTENT

Ж

PREFACE (PAGE I)
"Albert Einstein, is quoted as saying, 'Logic will get you from A to B. Imagination will take you everywhere'."

PROLOGUE (PAGE IX)
"Atoms, Solar Systems, Galaxies and the Universe."

PART 1 UNDERSTANDING GRAVITY
"An interpretation of the force of gravity, the beginning of our model."

CHAPTER ONE   (PAGE 1) - WITNESSING THE WARPING OF SPACE-TIME OR THE CONTAINER, GEL & SPHERE EXPERIMENT
"Our models need not be built of pure fact alone; it's our thought process behind our model that counts in our development."

CHAPTER TWO   (PAGE 11) - THE LAW OF ATTRACTION OR THE PUSH INTO THE FALL
"We are constantly falling in all directions at once. Let's look at the force or "push" of gravity, from yet another perspective."

CHAPTER THREE   (PAGE 19) - SUMMERY OF PART ONE OR OUR GRAVITY LAB MODEL
"Gravity is the force created by space-time against mass."

PART 2 THE GREAT EXPANSE
"Creating a model in the grandest of scales."

CHAPTER FOUR   (PAGE 29) - ORBITS AN IDEA OF HOW THEY WORK AND WHY OR, WHY SPACE AND TIME ARE LINKED AS ONE
"Space, plus time equals space-time."

CHAPTER FIVE   (PAGE 37) - THE CONSTANT SPEED OF LIGHT OR, LIGHT RINGS AND HUBBLE BUBBLE BITS
"A photon is a packet of energy. Tiny bundles of information. Ones and zeros, code which allows our adapted brains to interpret our surroundings."

CHAPTER SIX   (PAGE 47) - WHY CAN'T MATTER GO FASTER THEN LIGHT OR, CAN, WE, GO FASTER THEN LIGHT
"Matter (protons, neutrons and electrons etc.) or energy appears unable to "naturally" pass through space-time faster than the speed of light. Matter disperses space-time. It takes massive amounts of energy to disperse it quickly, relative to light."

CHAPTER SEVEN   (PAGE 51) - WHAT IS A BLACK HOLE OR WHAT IS 100% DENSITY
"You won't be sucked 'into' the black hole. The black hole is already full."

CHAPTER EIGHT   (PAGE   59) - HOW CAN THE FINITE BE INFINITE OR HOW CAN THERE BE NO CENTER
"Inside the universe there is only one true center and it's everywhere."

CHAPTER NINE   (PAGE   65) - WHAT WAS THE BIG BANG OR DID THE BIG BANG REALLY GIVE US SOMETHING FROM NOTHING
"The "Big Bang" was not a bang at all. It was a moment of a tremendous reaction. One, which dispersed an infinitely dense point of matter and by a means of processes, recreated the material, into the elements we see today."

CHAPTER TEN   (PAGE   71) - WHAT COULD OUR UNIVERSE LOOK LIKE OR INVERSE BUBBLE THEORY
Think back to our bubble on the oceans floor. As the tiny bubble is jettisoned upward, it begins its expansion. Upward, and outward. Growing in size with every meter it rises. The pressures gradually allowing the gas molecules to stretch, as the bubble moves towards lesser and lesser densities."

CHAPTER ELEVEN   (PAGE   75) - CONCEPTS OF DARK MATTER OR THE CRASHING OF WAVES
"In the freezing temperatures, the splash and mist of the waves collect as layers of ice, over the top of the great stones. Very quickly, the thickening layers of ice buildup, to encompass the rock. Increasing its appearance, the small boulders "grow" to much larger scales over the weeks to follow. Eventually collecting into what appears to be a single mass."

CHAPTER TWELVE (PAGE 79) - CONCEPTS OF DARK ENERGY OR WHY NO SPECIES MAY EVER SEE THE EDGE OF THE UNIVERSE
"In our model, it is not unlike our tiny bubble on the ocean floor."

CHAPTER THIRTEEN (PAGE 81) - IMAGINING THE MULTI-VERSE OR DO WE EXPAND INTO NOTHINGNESS
"Remember when we asked, "What's the multi-verse look like?" I still don't know, but let us attempt to imagine."

CHAPTER FOURTEEN (PAGE 85) - WHAT COULD HAVE CREATED THE MULTI-VERSE OR IS THERE MEANING WITHOUT A MAKER
"Kahunkna the sun, became so very jealous of Lunisica, his father the moon's beautiful brightness, that once whilst in a rage; Kahunkna cast a mighty star into Lunisica's heart, whose blood then poured out to form the earth."

EPILOGUE (PAGE 87) - WHAT IS MORE INSPIRING THAN BEING CREATED FROM A STAR OR LIFE IS A PAINFUL AND PRECIOUS OPPORTUNITY
"In an otherwise seemingly infinite collection of inanimate primordial elements, life formed, and has risen out from within our Universe. Ultimately, evolving to wonder enough to gaze back upon its self."

∞

PREFACE

Ψ

The following is an intellectual exercise used in an attempt to formulate a larger perspective of the cosmos, applying evidence, logic and imagination to lay the ground on which one is able to construct hypothetical tentative chains of thought. Doing so, an individual has the opportunity to formulate diverse models of various subjects, including the very universe of which they (or *we*) are a part of; thus providing potential explanations and viewpoints to subjects that may otherwise be unattainable or at times even undefinable.

Together, we shall endeavor to visualize a handful of the universe's known laws. Then continue further by speculating likely or at least possible solutions to other issues that as of yet, we are currently *only* able to speculate. Provisional or temporary concepts we are willing to amend as updated data is acquired or learned.

I wish to emphasize the importance of creating dedicated space within our minds to ponder life's grand questions, beginning in childhood and continuing through adulthood, as an ongoing and evolving lifelong experience; recording our altering understanding of this curious existence. It is one of the universe's most wondrous achievements: that a living creature is not only able to rise out of inanimate matter, but also to one day possess an imagination strong enough to question and attempt to learn great truths and solve great mysteries about itself.

A "Thought Experiment" is what some people might think of as a daydream. To me, it has become a method of focused mental exploration, during which one is able to ponder known testable facts and add seemingly impossible possibilities, within areas and ideas we have yet to fully

A CONCEPTION OF THE UNIVERSE

explore. A combination of established facts and speculative ideas, intended to open our minds to the enormous range of possibilities and subsequent truths, many of which we may regularly be apprehensive to consider. Although, I believe the marvels of reality are often far more inspiring and liberating, once recognized and absorbed, than most archaic mythologies and folklore.

In this essay, I will attempt to describe aspects of several thought experiments I have been postulating. We will be contemplating sizes and forces which, by all accounts, our human brains are scarcely prepared to immediately, or ever, completely understand.

I use over simplification in my postulations. Keep in mind that these ideas and metaphors are experimentations aimed at creating several possible, if not plausible, ways to structure or model the universe within your mind. These are large and complicated issues. I am forced to postulate them in the simplest terms that I myself can endeavor to comprehend.

"What's the multi-verse look like?"

Come on! How can anyone even begin to answer something like that? Would light even travel fast enough for there to be an outside view of the universe? Would there be some sort of reverse time dilation quantum craziness effect thingy, if one were capable of having a "god's eye view" of the multi-verse? Perhaps, allowing light, or somewhat of an equivalent to photons, to be seen?

"Maybe we can get back to that later."

It does not really matter. In our "lab" if we need to "see" what is around us, we may simply turn up the lights or travel closer. These and others are the simple forms of notions we are able to insert into our models. Since it is a "thought" experiment, we have the luxury to speculate different ideas in

EXPLORING THE UNIVERSE FROM WITHIN

order to gauge what may work in a larger picture, without the demands of being constantly rooted and restricted to absolute reality through every aspect of our model. We assemble our answers upon testable facts, then build upward with evolving postulations, to gain a greater perspective of knowledge and various possibilities within a concept.

You may call it "daydreaming" if you would like. I see it as my own personal laboratory. One from which, at will, I can explore, learn and share ideas.

"We will most likely wish to include some equations in our models, at least the shorter versions. We will probably get into a few things like '$E = MC^2$' or we may need to know how to measure the outside surface tension vs. the inside surface tension of a bubble. Who knows? Nevertheless, yeah, we would probably do well with a few sets of equations. We will just have to see if we need them.

Albert Einstein, is quoted as saying,
'Logic will get you from A to B.
Imagination, will take you everywhere'. "

As elegant and informative as a meticulously described and understood formula or calculation may be, we are not required to walk into every idea carrying our favorite abacus. Remember, we are not attempting to "prove" any hypotheses. We can at first simply postulate functional possibilities, then foster our hypothetical conceptions into arguments we are able to examine and evaluate; in order to pinpoint and cultivate any potential testable data within our models. For only when an idea is testable and repeatable, at least to some degree, does it formally earn the title of "Theory." Until then it remains conjecture, a hypothesis... purely exploration, wonder and deduction.

Once we are able to construct a compelling thought or model, one in which our towers of ideas appear to be

A CONCEPTION OF THE UNIVERSE

supported by other known towers of ideas...ideas proven by repeated testable results and the diligence of skepticism and scrutiny...only then are we able to actually analyze and quantify what exactly we may prove and what exactly we can still only call our *best guess*.

Ever since childhood *"presuming as most people,"* I have pondered many complexing questions. Occasionally the answers seem to come to us in black and white, good or bad, large or small... *2 + 2 = 4*. At other times, the answers may be fluid, ever changing or bewildering, sometimes at first glance seemingly unattainable, kind of like $E = MC^2$.

Yet, the most cryptic and detrimental of all, are the *answers* posed as convictions of truth. "Truth" only obtainable when taken at faith, or in other words, blind unquestioning devotion absent any real substantiation. Misrepresentations which offer the promise of illumination, only to leave one conciliated, and yet still in the dark. Ideas often handed down as time-honored practices, which may indeed lend an imagined personal comfort and sense of superiority to one's psyche, however, over and over again limit the extent of one's conceptual understanding, inducing guilt-ridden shame when questioned, and which are, more often than not, accompanied by inflexible and delusional justifications to immediately dismiss any opposing view regardless of how rational the arguments may be. Answers trusted at times simply because one is able to "feel it" inside themselves; nonetheless, in many cases, purely endured for fear of chastisement or, worse yet, retribution.

For a portion of my life, I naively assumed our universe to be a sphere shaped explosion, sparked by a celestial hand. It was my first actual notion of "The Big Bang." I believed it could have been a massive eruption, tearing a void in some higher dimensional space, creating a ball in which the universe is contained, expanding away from its center point. Outside of our universe was, in essence, too much to think about...or simply did not exist. I was convinced

EXPLORING THE UNIVERSE FROM WITHIN

it was just not worth contemplating, since it is obviously untestable and so far over my head. Who was I, to ponder such things?

Once, while whining about not having the resolve to ask an acquaintance a simple question, because I supposed it as intrusive; my great friend, sensei and mentor, Reese *'Old Man'* Shaner asked me, "Jeff, why would you presume to impose *your* limitations on others?" A short yet recognizably profound statement. One that constantly resonates throughout our internal assessments of not only ourselves and those around us, but also our actual perception of what we recognize as our existence.

More often than not, I tend to adopt his sage advice, as part of my own personal philosophy and this occurrence was no exception.

"Thank you, Sir."

In turn, I have asked myself, "Why should I allow another's credulous limitations to impede on my attempts to raise questions and find legitimate solutions, both vital to my nature and sense of being?" Uncertainties which are frequently unanswerable in solitude, and are often significantly larger than one's own personal experiences.

At the very worst, our postulations will be wrong, and we will be forced to correct them as additional evidence is presented or learned. However, at best, we may stumble upon a possible glimpse of a greater understanding in a subject important to us. Concepts to which we otherwise may have been oblivious, and at times hesitant to scrutinize or, in some cases, even dare to imagine.

There are so many questions that we will take with us to the grave. The idea of never knowing the answers to some things can, at times, rack us with fraught. Some days I feel such empathy for those of the past who so desperately

A CONCEPTION OF THE UNIVERSE

struggled to learn the world around them, only to never uncover their truths, leaving this world confused, misled and in wonder. Then again is it not, very often, the *wonder* that makes life beautiful and exciting? The overawing potential of unending possibilities?

Think of the great Galileo's reaction to images we see from the Hubble Space Telescope, or perhaps the Mars Rovers. The joy he would have received if we were to whisper in his ear the possibilities of Jupiter's distant moon Europa, with its vast salt-water ocean hidden deep under its frozen surface. Consider his awe at the simplest concepts of the Big Bang.

Now next imagine all of the future knowledge and exploration we will miss out on after our time here has ended. There are moments when I feel the loss so deeply it can be maddening. It is impossible for me to not be envious of our future generations. However, I am so very inspired for, and excited by, the potentially positive and emboldening prospects possible for our world. The achievements, pleasures, contentment and adventures the future may bring to our kin, although, of course, they too will have their own challenges and struggles.

Barring the crippling or, worse yet, the extinction of our species, the future rate of our advancement in science over the next 100 (or imagine even over 10,000) years will absolutely grow in orders of compounding magnitude. It is certainly reasonable to hope, and even expect, to one day be able to answer our most challenging questions with confidence. It shall simply be a matter of time.

EXPLORING THE UNIVERSE FROM WITHIN

"We will never completely comprehend all that we wish we could. It is true. Yet would we allow that line of reasoning to prohibit us from exploring new ideas and wondrous possibilities? No!
We may not be able to solve it all, but we can still ask some big questions."

A CONCEPTION OF THE UNIVERSE

The techniques of formulating well-devised thought experiments may help to enable us in constructing a variety of models, in this case of the universe, which we are able to explore at will.

Within a thought experiment we have access to an infinite size laboratory, and it is free to use! We also have available to us a whole imagination's worth of technology at our disposal. These tools offer us very effective opportunities to visualize images and gain new perspectives of nearly any concept we can imagine. Our models do not assert themselves as being perfect; they will often change and ultimately *should* change over time.

I suggest we will be developing models which are not unlike the proverbial house of cards. Building layer by layer starting with what we *think* we know, or more precisely, how we *think* we understand it to be.

When two ideas collide and do not fit together, then one idea, or both, tumble down, and we are forced to reexamine our positions. In doing so even faulty models can conceivably open our minds to new ways of understanding...sometimes comprehending concepts we did not even realize we didn't know.

"And in the end, we must always ask ourselves, how has our understanding of the evidence changed? What more can we add to our models? Most importantly, is it yet time to rethink them all?"

PROLOGUE

∞

"Atoms, Solar Systems, Galaxies and the Universe."

Matter consists of Atoms, which are sub-microscopic collections of various particles that form into their own small systems. The Nucleus, which is the center of an atom, consists of sub-atomic particles called Protons and Neutrons; tiny packets of energy formed from even smaller particles called gluons and quarks (it gets a bit complex at deeper depths of size than that, yet is unquestionably worth studying further). Protons have a positive electrical charge, while Neutrons are in a neutral or uncharged state. The nucleus is orbited by negatively charged particles called Electrons, which form an energetic field surrounding the nucleus.

The atom is its own extremely miniscule spherical system of incredibly small particles with an unbelievably vast amount of space in-between the nucleus and surrounding electrons. Atoms can remain pure, collecting to form elements such as gold, carbon or uranium. They are also able to combine together with different forms of atoms to create a collection called a molecule.

Water is a perfect example. H^2O, or two atoms of hydrogen and one atom of oxygen, combine to form a water molecule. Various combinations of atoms form the diverse compounds of matter that in turn form us, and every other bit of matter we are able to see around us, constructing the entire known universe.

A *Solar System* is an enormous collection of atoms! These systems consist of at least one main item, a super massive sphere of various elements in a gaseous state (mostly

A CONCEPTION OF THE UNIVERSE

hydrogen and helium, the most simply formed and abundant elements in the universe) which are condensed so tightly that nuclear fusion is ignited within the center of the sphere. The immense power of the fusion reaction within sets off an amazing balance of an incredibly powerful outward exploding force vs. the absolute crushing strength of gravity's push into the massive collection of atoms.

This results in what we nonchalantly refer to as a *star*, an astonishingly large ball of luminous gases, with a release of energy that essentially creates an embryotic shell of protection from many of the dangers of interstellar space, sheltering its planets, or its *system*, against copious amounts of radiation traveling from innumerable points of extremely hazardous occurrences, along with interstellar winds, rogue objects, and many additional forms of supremely powerful energy events.

The star forms within the center of an enormous gas cloud. The shockwave from the initial rupture of fusion blasts away and condenses the remaining outer gases; eventually collecting to combine into and form clumps of solid matter such as planets, moons, and various debris (which includes you and me).

A solar system (which I assume is coined after our own sun, Sol) consists of one or more stars surrounded by the remaining elements of matter, which evolve to form the orbiting material contained within the protective field established by the fusion event.

A *Galaxy* is an even larger collection of matter, one that contains millions of stars and in most cases literally hundreds of billions of star systems. Galaxies too, form out of extremely huge clouds of collapsing elemental gases; creating globs of stars and dust, often with time, forming into swirling pinwheel discs.

What is known as the "Hubble Deep Field," is an image taken of thousands of such galaxies in a pinprick size point of our night sky. Experts have collected data and images showing we are surrounded by *hundreds of billions* of these

EXPLORING THE UNIVERSE FROM WITHIN

galaxies, all much like our very own Milkyway galaxy, with immense distances of empty space in-between them.

Our *Universe* is an exceptionally vast collection of (at the very least) hundreds of billions of galaxies... and those are only the ones we are currently able to detect. It is likely the total universe consists of trillions of galaxies, each containing millions and, more often, many billions of stars; every star potentially surrounded by several orbiting planets, all consisting of trillions, upon trillions, upon trillions of atoms!

Due to the force called "Dark Energy," the universe is an *expanding* assembly of galaxies, held together within space-time by the incredible force of gravity. One could continue to speculate that our universe is also only one of many. One of an infinite number of universes, which emerge and are eventually extinguished, in a revolving circle of timeless creation; possibly carried out over trillions of millennia.

"Now that we covered some of the basics, let's have some fun!"

PART ONE

Ψ

UNDERSTANDING GRAVITY

CHAPTER ONE

§

WITNESSING THE WARPING OF SPACE-TIME

OR

THE CONTAINER, GEL & SPHERE EXPERIMENT

"Our models need not be built solely of pure fact alone. It's the thought processes behind our models which guide our development."

"We'll now begin our first experiment."

Step one. We will commence by creating a virtual laboratory within our minds. Here, we have unlimited resources at our disposal. Next, we need to simulate the infinite vacuum of space. We will also need to conjure up a one-meter cubed, transparent, and unbreakable at any size, box. We will call it our container, container one.

Now, let us completely fill the container with our own transparent, but ever so slightly green, very fluid, cosmic super gel. Let us think of it as a clear, seemingly indissoluble slime, which has a very elastic memory able to stretch or compress infinitely, precisely matching the volume of any vacuum or space it is contained within. When our container has been filled, we can seal the lid at the average surrounding pressure.

The last crucial piece to step one is a tiny little metallic sphere, a grain of super steel, which we can say weighs approximately 0.1 gram, at a density equivalent to (x). Next,

A CONCEPTION OF THE UNIVERSE

we will need to insert - No! Let's *teleport* - our grain of super steel into the gel to the very center of container one.

We will call this point of our experiment *step two*. Let's zap our box with a fancy little ray gun stored within our brain's arsenal of hypothetical new technology. With its beam, we can increase the size of our container to a point that the interior space equals 100 meters cubed. At the same time, we will allow our gel to expand evenly throughout the container.

Remember we stretched the container out but kept it sealed. Now the pressure inside has what? Increased! Rather, in a negative direction. This increased vacuum has pulled our super gel along with the container's walls, matching the full interior volume of the container. We did not need to add more gel, just gave our gel more room to relax.

Each particle of gel (say, particle A) is surrounded by the very same particles of gel (we will call B, C, D, E, F and G) which it had been surrounded by before the gel was expanded. The amount of gel mass, atom for atom, or maybe better thought of as unit for unit, has not changed. Another thing we did not change in the container is our sphere, our tiny grain of metallic super steel. It is still tiny and it's dead center in the middle of the container, suspended within the gel.

Step three. Let us zap our grain of steel. We will expand the sphere's volume but, unlike the gel, this time we will also increase its mass, say by a factor of 100,000 times. There it sits, right in the middle of our container. We have every view of it possible at our disposal. As we turn up our lighting and gaze towards our sphere, what do we see?

What was a pinprick size mark in our gel has now grown to look like a large, shiny cannonball. As our lighting shines through the container, it passes through the gel surrounding the sphere. Here we are able to visualize the compression made by the now enlarged sphere within the gel.

Not unlike the warped view we see when we look down at half our legs submerged in water (where they appear

EXPLORING THE UNIVERSE FROM WITHIN

shifted from their "true" location), so looks our lighting as it is passes through the compressed gel.

The gel is compressed against the sphere with the most pressure and distortion (say, 'gel curvature' representing 'gravitational lensing') nearest to the sphere. A compression in the gel that lessens more and more as we look in any direction away from the sphere until the distortion over distance dwindles and appears unnoticeable several meters away. The closer to the sphere, the more noticeably compressed the gel has become.

"Can you visualize it within your mind's eye? To me, it is the beginning of a glorious image. Let's press our mental "Save Button" and we'll return to "container one" in a moment."

Let us now start again back at the beginning of *step three*. We have our next container and our expanded gel. In container two, we are about to enlarge our grain size metallic sphere again. When we enlarge the sphere, we will again expand its size or volume by a factor of 100,000, but this time we will hardly increase its mass.

"We haven't done the math, but..."

We will assume for now that our grain size sphere has approximately one billion or 10^9 atoms, that make up its original grain size mass. This time, as we expand the sphere, we will only increase the sphere's mass by ten thousand fold. Mass which is now dispersed throughout our now cannonball size sphere.

What do you suppose we will see now when we turn up our lighting? Let us zoom in to a microscopic level. We observe our gel quickly seeping back into the sphere between the voids left over from the expansion of the grain's now *measly* 10,000 billion atoms. Pushing back into voids opened as the result of the atoms traveling apart from one another, an effect produced by the sphere's density being lowered, and in turn causing the space between the atoms to increase.

The gel pushes at the surface of the sphere, from all directions. The gel travels down as deep as it can go towards the center of mass, pushing to return to its "wanted" location. Each particle has its own "force of return"; the magnitude of the force is compounded by the additional forces added by every other displaced particle of our gel. That is to say, the other "units" behind the furthest most forward displaced unit.

So, within our model, the effects of the varying forces are governed and compounded by each additional particle of gel (or "unit" of space-time) displaced, each unit wishing to return to its preferred location in the vacuum. The force comes in steady waves of magnitude. A strong force if the space-time unit is greatly distorted or greatly removed from its preferred location. A weaker force if the space-time unit is less distorted or less removed from its preferred location.

The force diminishes more and more, relative to the distance from mass and the degree of displacement per space-time unit. It is a push from behind to assist the lead or foremost removed unit of space-time, which in turn aids each unit to gradually return to its preferred location. Assistance that aids the most displaced units to struggle their way through the density of matter.

"The harder space-time has to push to return to its 'wanted' location, the denser it packs the matter in its way.
Higher density equals harder push of return.
Harder the push, the denser the atoms are compacted.
A circling struggle of opposing forces.
When I imagine a "unit" of space-time, I endeavor to visualize it as a 'Planck length'.
That's about 636.30 times 10 to the $^{-36}$ of an inch. It's small, really, really small."

EXPLORING THE UNIVERSE FROM WITHIN

Let's zoom back out. As our lights shine through the gel surrounding the sphere, we can still see distortion, but if we were to measure the pressure of the gel against the sphere at the spheres surface, I would wager that its pressure or, "tension," would be much less than that of our original model. Container One, where we increased not only the volume significantly, but also and more importantly, the mass, of the sphere by a whopping one hundred thousand times.

As we look at the gel surrounding the sphere in container two, we see the distortion is noticeably less than in container one. The distortion would appear to travel a lesser distance from the sphere in container two (in any given direction), than it did in our original *heavy mass* sphere expansion model in container one. Namely because of the obvious differences in surface penetration of the gel between the two spheres.

If next we were to shrink ourselves to a comfortably microscopic size, then teleport ourselves to the surface of the spheres in both container one and then container two, we could assume we'd easily feel, or better yet, be able to measure, the pressure differences pushing down upon us between the two containers.

I say, "down upon us," which would be the greatest direction of force being applied to us. However, the gel would also be pushing at us less noticeably in every other direction as well. In essence, squeezing in on us as it packs around and in-between our atoms as much as it possibly can.

The gel, or "space-time," wishes to penetrate the location we consume, but with a lesser noticeable force due to its shortened distance from its preferred location, and as well by the amount of equilibrium the density of our mass allows. The greatest force would still be directed at the center of the largest mass, and its highest density, in this case, the sphere below our feet.

Chapter One | 5

A CONCEPTION OF THE UNIVERSE

We would not only detect the difference in pressure rates per sphere based on the different degrees of penetration by the gel into the various densities of the spheres, but also the deeper we traveled into each sphere. The compounding force increases more and more the closer we travel towards the center of the spheres (or the area of the mass's highest density), until the point at which the gel is crammed so tight between the atoms, it takes too much energy to travel or squeeze in any further, proportioned to volume of space-time displaced.

One more step. Let us teleport our lab, and the entire experiment, to the surface of the earth. Next, we can allow the pressure in our containers to equalize to the surrounding outside pressure. We will still keep our boxes and gel expanded. One more zap from our custom ray gun and we will clear the upper half of the gel away, exposing precisely one-half of each sphere.

Now, with the aid of the gravity on Earth, we can see how the density of each sphere warps the newly exposed surface plane of the gel. It is the classic ball on a rubber sheet experiment, sagging more and more until it comes to the curve of the sphere, where it then dimples and bends to match the push of the sphere's mass. The gel is deflected by the sphere's *intrusion* into the gel's preferred location, sagging less and less the further from the sphere we look.

Now, imagine the difference between the surface planes of both containers. The denser sphere pushes down harder on the gel than the less densely packed sphere. The sphere with the lesser density would appear, and surely could be measured, as having more "buoyancy" upon the plane of the gel than the denser sphere. Less mass, less applied force to the gel.

EXPLORING THE UNIVERSE FROM WITHIN

"What if we observed both of our models upside-down? We will rotate our views, not the containers themselves. Inverted, it now appears the tension between the surface of the sphere and the surface of the gel is being caused by the compression of the gel pushing down upon the spheres.
In the vacuum of space, space-time or the "gel" pushes at matter. Not from one direction, as in the 'ball and rubber sheet' experiment, but from all directions at once."

(We will come back to that statement and cover it more thoroughly in the next experiment.)

IN REVIEW

∞

When our models were in the vacuum of space, we could see the enlarged spheres and how they displaced the gel units from their original location, appearing to tear, but actually only displacing, the units away from their "wanted" or "predetermined" locations, creating not really a hole, but more of a *displacement* of our gel.

"Or, in deconstructing our experiment's metaphor: creating a temporary cavity in the very crux of space-time itself."

Space-time seems to desperately want to return to its "equilibrium," to once again fill the location which matter now possesses. It will do so with a steady pace and at a diligent compounding pressure, creating more and more pressure or "push" back towards the center of the mass. It will plow itself into every nook and cranny possible until it reaches its maximum force against the various densities of mass...or it reaches its destination, crushing its way to the center, back towards its "wanted" location.

The farther any one particular unit of space-time is distorted by mass from its wanted location, the greater the force space-time has, against the mass, to return. Individually the space-time unit has little strength, as long as it is not far from its wanted location. As you distort it from its mark more and more, the push against you will become stronger and stronger until you run out of enough energy to distort the space-time unit any further.

That is why, as we may imagine, traveling at light speed would consume a nearly infinitely large amount of energy. The push through space-time by your velocity creates a warping of space-time in front of you. Space-time eventually resists your push, seemingly canceling your increased momentum.

The effect is so weak on small scales of density or velocity, it can be difficult to visualize. Oppositely so, would be the immense power of the space-time unit as it reaches towards the scales of infinity against the resistance of mass, either in density or in velocity.

Visualize how the effect of pushing an object through space-time at extremely high velocities would push back at the object, creating in effect, the appearance of squashing the traveling mass to a shorter length as it plows through the gel of space-time, warping light and time. Generating a stronger and stronger "shockwave" in front of and behind the object, as it increases its force against the "gel" of space-time.

From a distant external point of reference, the accelerated force bends the space-time between the mass and the observer, warping their view of the mass. The curved space acts as a focal lens, making the mass's appearance seem shorter from an outsider's perspective.

The light we would perceive, if we were to take super light-speed action photos, would show the front of the mass being squashed, condensed against space-time as it travels,

EXPLORING THE UNIVERSE FROM WITHIN

distorting its shape. The same curved effect offers us a view of the rear of the object appearing squashed as well, however towards the direction of the mass's forward velocity.

From within the matter, we could propose the view of the mass would not change. Perhaps there would not be enough distance in the field of view to show us any visible signs of distortion. Then again, one could also postulate a brain may have difficulty in its interpretation of its collected photons.

"I can certainly see how traveling at light-speed could make you a bit queasy...at least if you were looking out the window."

End of Experiment.

§

Note: *"Ponder on this, too. At high enough densities, when the gel (or the "push" of space-time's gravity) comes close to an infinite amount of negative repelling force against that of mass's density, and as the force of return reaches levels of density close to or up to a magnitude of infinity, we can expect a "black hole" to form. A galactic battle of wills between two magnificent forces. What would you add to our model? What did we miss?"*

CHAPTER TWO

§

THE LAW OF ATTRACTION

OR

THE PUSH INTO THE FALL

"We are constantly falling in all directions at once. Let's look at the force or "push" of gravity from yet another perspective."

To further visualize gravity and its "pull," I suggest we conduct another experiment. Let's get back to the Lab. Imagine one of those plasma orbs you can buy at a novelty shop. We turn it on and see the center globe light up with bolts of plasma. From purples and pinks to yellows and whites, our very own theoretical lightning bolts, no atmosphere required. Let us remove the outer globe and pretend it still works without killing us. We are now watching its electrical tentacles reaching out in all directions.

When the outer globe was still present, we could touch the glass and see how the charge inside is attracted to our finger. (Being grounded perhaps? I assume so. It does not really matter for this model.). As we are set up, our plasma ball is sitting all by itself floating in a vacuum and we have control over every possible viewpoint.

A CONCEPTION OF THE UNIVERSE

As we watch the plasma tentacles reach out into nothingness, we see in the distance, globe two, another equal size plasma ball without its outer globe reaching out with its plasma tentacles...and it appears to be traveling closer. Not directly at globe one, but globe two will definitely pass nearby. As it comes closer, we watch, and can already guess at what we are about to see.

At 100 yards, it appears that neither globe has any idea the other exists. At 50 yards, they both seem to be changing their color pattern, glowing a slightly deeper purple on the surfaces facing each other. Now, at 30 yards, we can obviously see the pair have some connection, some attraction or "awareness" of each other.

At about 20 yards, we see the first faint bolt of plasma discharge across the expanse. It appeared first to come from globe one. As the bolt shot out from the globe, it traveled approximately half the distance to globe two. In an instant, you could see globe two illuminate and flare out in the direction of globe one. Globe two, while still waving its tentacles in all directions, then strikes out a bolt aimed to meet the burst from globe one.

We see the faintest of connections in the outstretched tentacles as they meet. Where the charges seemed to lightly graze, they ever so silently, yet very energetically, collide, and grab each other in a tangling grip, if only for an instant. As the two orbs get closer still, the tension steadily builds. The charge flashes out from globe two. Globe one meets globe two's bolt right in the middle.

And CRACK!

"That's right; we can add sound to our vacuum if we want. It won't really factor into the final results of our experiment. This is only in our mind's eye after all."

EXPLORING THE UNIVERSE FROM WITHIN

To simulate the force of gravity further still, let us add a magnetic charge to our globes. We see globe one, which up until now floated stagnantly in the expanse, is now listing ever so slightly towards globe two.

As the two globes near their apparent closest trajectory, it becomes clear that globe one is building momentum and is moving towards globe two. All the while, globe two is shifting its trajectory in favor of one's location, due to our magnetically synthesized gravitational "pull."

Our globes each still reach their tentacles out in all directions. We can see the mutual attraction they have for each other indicated by the growing charges reaching out into the direction of one another. We also see the attraction of the momentum created by our magnetic force.

We see globe two start to careen towards the direction of globe one. Now they seem to be on track to just miss each other in their passing. As we watch, we see two moving towards one at a fair rate of speed, and globe one's velocity is starting to increase more and more as well.

Soon, they will finally reach their closest point relative to each other. Globe two shoots past globe one. Immediately we see one change its trajectory, curve, and continue its momentum towards two.

At the same time, from our perspective, we see globe two's momentum slow slightly and bank, in order to circle globe one. With what appears as another small tug, one swings its momentum and continues to follow two's path. We see the dance that a pair of gravitationally locked objects produce.

They will swing around and seem to tug each other until they collide, fly apart and eventually settle into a mutual swirling dance. All the while, their bolts of plasma dance in unison with the magnetic pull coming together in a blazing twirl.

Our magnetic "pull" actually represents a "Pushing Force" towards a massive object. I want us to see matter as constantly falling in space-time in every direction all at once.

A CONCEPTION OF THE UNIVERSE

*"Yeah, it sounds peculiar at first.
But I am going to say it again at least a dozen more times so,
you'll have a chance to get used to the idea."*

In our model, gravity is more like a push from behind towards a greater center of mass, than a pull given from the mass, as one might think at first glance. Here we see mass as falling in every direction in space-time all at the same moment. That is fine... for an object impossibly stagnant in the vacuum of space-time.

Add velocity to the mix. Give the mass a shove, and now the object moving through space-time is not only falling in space-time in every direction at once but, furthermore, it has its forward momentum. The object is falling in all directions at once, but falls with its added momentum faster in one direction, being pulled, or "pushed" as I like to think of it, towards any and all other masses. The degree of which varies, depending at minimum on the relative distance between objects and the mass of the objects involved.

With the plasma tentacles reaching out in all directions at the same time, we get an image of the "attraction" of gravity. With our magnetism, we also see an image of gravity, as the two objects magnetically pull towards each other. The pulling force of our magnets towards each other represents the returning force made by space-time, produced from the displacement of space-time by matter.

While this metaphor may allow us to visualize gravity's pull and reach, it also and, moreover, represents an exact *opposite* view of what is really happening. Our magnetic pull comes from our globes reaching through space-time. In reality, the force is manifesting from space-time, not reaching from, but pushing back at, the globe.

The same goes for our plasma tentacles. In our metaphor they came from the ball...that image is actually the exact *opposite* of what is happening around us. The weakest

EXPLORING THE UNIVERSE FROM WITHIN

force of push from space-time is farthest away, greatest forces of push are generated by space-time nearest the center of the mass, in the direction of the highest level of space-time displacement and mass density.

The "push" of space-time against an object into another object that has connected gravitationally with the first, begins the objects' assent into the others' gravity well. The objects seem as if to fall towards each other in an accelerated rate.

The closer the objects come to one another, the faster they fall into space-time's "push" towards one another. Matter then receives an extra compounding "push" from space-time as the object crosses a field of space-time attempting to return "home" to its wanted location, endeavoring to maintain its natural symmetry.

This is the "gravity well" that makes an object fall into another object's direction at a greater rate of force than the average "push" in all directions, or what *would be* the average uniformed push created by matter displacing space-time, if the matter was at an impossibly stagnant position.

If we were a very large center of mass, we would feel that push equally in every direction all at once, squishing us into the smallest surface space possible; a sphere, if our mass, volume, density and element structure allow. Space-time will also crush us to the densest point possible in which our mass size and element structure will allow.

A CONCEPTION OF THE UNIVERSE

IN REVIEW

∞

I want us to focus on this point: *We are constantly falling in all directions at once.* All mass, is constantly falling in all directions at once. If it were not for other objects of mass in the universe, we would be falling at the exact same rate in all directions at once, appearing as if we were stagnant.

The "unit's" attempt to return to the symmetry or equilibrium of space-time's "wanted" location may not seem like much at first. However, remember, space-time has all of itself, a seemingly infinite self, to assist in some degree, "encouraging" that small portion of space-time to achieve its goal and return "home."

Again, the force of gravity pushes at us from all directions at once attempting to recover its loss of location. Remember, mass has displaced space-time. Space-time, essentially "wants" or "needs" to return to an equilibrium, penetrating mass relative to its volume size and density range.

Our resistance to let space-time pass into and through us as much as it possibly wants, is what regulates the "pushing" force upon us, creating what is measured as weight and/or density.

If we imagine you as very dense (*don't take it personally*) we could expect the harder space-time has to push on you trying to dig its way between and into your atoms, the greater the force is applied to your density level. At high enough densities, the push becomes overwhelming, initiating the fusion of atoms and unleashing space-time's ferocity.

End of Experiment.

§

Note: *"How would our model appear if we dialed up one of the globes' pull and lowered the other? How would that affect the orbits of each? How would they move if we sent in more globes of varying energy levels? What else could we do with this model?"*

CHAPTER THREE

§

SUMMERY OF PART ONE

OR

OUR GRAVITY LAB MODEL

"Gravity is the force created by space-time, against mass."

We proposed in our models that gravity is not generated from a force produced from mass, such as the classic attractive force model the magnet. Rather, it is the reaction of mass's displacement of space-time...a returning force or push. As mass accumulates, it replaces space-time in that given region...gravity is the returning force, the "push," of space-time's preferred symmetry.

We should consider every unit of space-time as having its own particular "wanted location" within the matrix of the universe, every unit designated to each location along a framework we perceive as space-time. It can shrink, stretch, bend, warp, and twist, but it "demands" to return to its "sense" of balance.

A CONCEPTION OF THE UNIVERSE

If we replace its location with matter, the mass displaces equal amounts of space-time, causing the space-time to warp and compress; pushing back towards its wanted or original location. The denser the mass, the greater the "push-back," or *gravity,* is recorded.

Space-time will fill any void left, down to the smallest areas between and within, any atoms possible. As matter accumulates, space-time is driven out of the area it once occupied; the force, then, is not given by mass, but by the vacuum of space-time, attempting to push back into the area the mass now occupies, struggling to regain its prior position of equilibrium.

The more mass, the more force is exerted, crushing evenly from all directions at once, creating the densest mass in the center. The denser the mass, the more space-time is squeezed out from any given area...a revolving circle of forces generating an even greater inward push or force against matter; causing higher densities, along with warped space-time encompassing its invader.

Once again, gravity is not a force *pulling* you *down*, it is a force *pushing* you *in*. The denser the mass, the farther space-time is pushed away from the mass's center point, creating a stronger gravitational field. The less dense the mass, the more room space-time has to fill; this generates a proportional balancing effect of less force per volume or, less *weight*.

∞

When we think of gravity's diminishing strength over any specific distance, we are able to compare its dwindling force, to the dispersal of light, as light, too, disseminates in the same manner as gravity. I find it is easiest to visualize by starting from the obvious and most common source of light we witness, a sphere shaped star. When we think of the surface

EXPLORING THE UNIVERSE FROM WITHIN

area of a sphere at any given moment, the sphere itself is finite at any exact moment in time. But, as time moves forward, the light that travels away in all directions is a seemingly ever-growing globe of radiance.

On to the Lab! We shall start in on this next experiment with a perfectly flawless sphere. We may never find one in nature, but here, in the laboratory, we are allowed brief moments of perfection. I recommend we create our sphere from a thin layer of unbreakable, lightly frosted crystal, with a smooth and polished outside surface, a globe of absolute precision. We can place our ball in a never-ending expanse.

The surface area on our new sphere will be forever unchanging; it will hold its exact size for eternity. We can easily calculate the surface area, since we are in control of its size. Specifically, we know its *radius*. The radius is a line from the exact center of the sphere directly to an outside surface point, half its total distance across or, half its *diameter*. We can give it a number or, *value*, but it really does not matter what its size; the rules do not change, we can simply say the value is (r). With this value we can use a simple formula to calculate the total surface area; four, times *pi* (commonly estimated as 3.14), times the radius, squared. Therefore, when written out it looks like **$4\pi r^2$**; fairly straightforward mathematics, but can be applied to many applications.

For the sake of keeping things simplistic and easy to visualize, we will give the sphere a value; we can designate its size as an unexceptional six meters across. Six meters in diameter means the radius is, obviously, three meters. So, let's work this one out quickly. Four, multiplied by pi (or 3.14), equals 12.56, take that total and multiply the sum by the radius squared or, in our case, three times three. Therefore, the total surface area of our sphere is 12.56 multiplied by 9, which equals 113.04 square meters.

I will place an impossibly small microscopic anomaly, a currently inactive star, within the exact center of our orb. It

A CONCEPTION OF THE UNIVERSE

will provide us with all the light we will need to illuminate our sphere. Throw the switch, ignite our little fusion ball, and we will instantly freeze the light, or photons, precisely at the sphere's surface.

In this model, we will think of light and gravity as one. We are using the light to visualize gravity's force, but both light and gravity inversely lose or gain strength relative to the distance from their source. More precisely, what is known as the *Inverse Square Law*, explains how the intensity of gravity (also light, sound waves, radiation and even electricity) spreads out over any particular distance away from its source. The intensity of any given force, twice as far from its source, spreads out across four times the area. That is to say, the value diminishes to one quarter of its original intensity, relevant to the same surface area, twice the distance from the source as was previously recorded.

We can monitor our now lit orb frozen in time before us. We will record the amount of light exiting any given one square centimeter of surface area, one photon thick.

"This might make it confusing, but let's give it a go."

We can give our escaping energy a value...since we are presently contemplating gravity, we can label its value as (G). The value of (G) we can designate as (x) amount of units, for now we can say (x) equals 32 units; we will assign each unit as containing one layer of one trillion photon particles, one particle thick. We recorded that measurement at the surface of the sphere, three meters from the center; we can call that distance point (a).

Slowly, start time. As (G) departs point (a), it travels outwards towards point (b). Point (b), is twice the distance from the very center of the sphere as was point (a). We went from three meters, to six meters in distance away from our miniscule star. Here at point (b) we can again stop time. If we erase all other escaping photons other than our original

EXPLORING THE UNIVERSE FROM WITHIN

squared centimeter; we see a square beam of light extending from point (a) to point (b). The size of (G) has grown over the distance to a size of four times its original designation, but it still contains only 32 trillion photons. Our frozen beam is now two centimeters wide, by two centimeters tall, still only one photon thick.

At point (a), we had a concentrated 32 trillion photons, thirty-two values of (x). At point (b), we still have 32 trillion photons, but they are now spread out over a plane four times the size of point (a). That leaves us now with only 8 trillion photons per square centimeter. Next, on to point (c). Point (c) is twice the distance from our little star as was point (b). That means, twelve meters away from the center of the sphere. Our square beam of light at point (c), has now stretched to four centimeters wide, by four centimeters tall. Our 32 trillion photons have now spread across a plane of sixteen square centimeters, leaving us with 2 trillion photons per square centimeter. If we allow the beam to travel to point (d), twenty-four meters away, (G) will have grown to sixty-four square centimeters, leaving us with only 500 billion photons per square centimeter. So on and so on.

It goes on like that forever. Gravity's force continues to diminish over distance, just as the photons spread their energy, dimmer and dimmer as the distance mounts, until the force is so diluted it becomes lost in a sea of distant, stronger forces, overtaking the given position.

"I illustrate the image in my mind, somewhat as throwing a pebble into a wave...the pebble's faint ripples are quickly consumed by the wave's own underlining force."

§

A CONCEPTION OF THE UNIVERSE

***Note:** "We should consider every unit of space-time as having its own particular "wanted location" within the matrix of the universe. Every unit "designated" to each location along a framework we perceive as space-time. However, a unit of space-time can exchange positions with another unit of space-time and still be content. Perhaps not completely unlike a displaced water molecule deep within a sea. It is, for lack of a better term, a "collective." A cooperative, which will stand united as one. It will shrink, stretch, bend, warp, twist, and expand, but it "demands" to return to an equilibrium, whenever opposed by mass.*

When we replace a given location of space-time with mass, the mass does not fill space. Rather, it displaces space-time, causing it to warp and compress into what we can observe as gravitational distortions. The denser the mass, the greater the "push back" of gravity recorded. Space-time's returning force...due to the density of the material and the volume of displaced space-time from within any given location, right down to between and within the atoms themselves. Gravity's force diminishes, measured as an outward growing sphere, weaker in proportion to its distance from mass, evenly distributed in a widening orb of fading intensity. Inversely, its power compounds rather quickly as one travels towards a massive object."

PART TWO

Ψ

THE GREAT EXPANSE

CHAPTER FOUR

§

ORBITS, THOUGHTS OF HOW THEY WORK, AND WHY

OR

WHY SPACE AND TIME ARE LINKED AS ONE

"Space plus time equals space-time."

What is *space-time* exactly? Most simply put, space-time is comprised of the *four* dimensions with which we are most familiar. Three-dimensional *space* being height, width and length, and the fourth dimension being, of course, *time*. Together they make up our four-dimensional space-time.

If you move an object through space it will need to move through time, but moving through time does not require you to move through space. Although everything within the vacuum of space is continuously falling, it is relevant only to one's perception.

"A galaxy is an incredible mass of swirling madness, a dazzling array of seemingly timeless orbits."

A CONCEPTION OF THE UNIVERSE

In the Lab, we are able to stare down upon the Milkyway galaxy, then apply our "god's eye view" to the enormous cluster of matter before us. We are able to see past the immense amounts of "space-dust," clouds of coalescing elements, which would normally obscure our view. We are able to survey the stars below and catch sight of hundreds of billions of objects, trillions of points of movement. Stars and planets in their orbits, swirling drifting and wobbling elliptical rotations; asteroid fields, comets and glowing nebula, all mixed with great rates of varying velocities and grand, incredibly powerful collisions. At the time scale of three Earth years per second, we are able to recognize the constantly developing and evolving, repeated patterns. A morphing and complexing clockwork which essentially emulates life itself. An absolute flurry of exquisite elliptical arrangements, whirling around and around, not unlike a flourishing swarm of gnats caught in drifting circles, chasing other sparks of light. Constantly falling in elegant patterns orbiting their home star, as the star itself drifts away, orbiting other far distant stars and clusters of matter.

Nothing is ever truly still. From our perspective, as we speed up our journey through time (say, up to 100 years per second), we see a blurring whirl of movement. At 1,000, if we pull back our view, we can even witness the slow orbits of galaxies themselves, a few caught in each other's gravitational push towards the other, a marvelous dance of mass and photons.

At 20,000 light years per second, we are able to witness galaxies collide across the expanse; they swirl around, back and forth, until they eventually congeal and blend into one. We also see the spreading effect of expansion made clear...on a whole, as we gaze out, the vast majority of the galaxies are drifting apart.

EXPLORING THE UNIVERSE FROM WITHIN

"Thus far in this experiment, we've been focusing mainly on two points. Space-time and orbits. We can define space-time pretty quickly, so let's cover that first."

As we peer across the universe's great expanse, we are in clear view of four-dimensional space. This is "true" 3D, and then some, for staring through space is literally staring through time.

Even with our lab's most superior view, we cannot help but see photons that have been traveling for countless millennia. The closer the galaxies are to us, the more mature they appear as we gaze deeper in any direction. What we are doing, is peering back in time. Hundreds of thousands of years, millions of years, and, as we look out farther, our eyes absorb photons that have traveled through space-time for billions of years. We see galaxies as they first began to cluster, an actual visual history of our universe's development. Not a guess...there is no *wishing* we knew...it's right there before our eyes. Amazing!

The universe is so very vast, that one simply cannot look to the skies and truly see the present moment; only what we *perceive* as the present view, at a particular moment and location. Light is very fast, but it still takes time to travel through space, just as mass takes time to move from one location to the next. We easily see every day that movement takes time.

The glass on the table can maintain its current location quite simply, it just sits there going nowhere, yet it is moving is it not? Maybe not through space relative to the table, but it is still moving through time. If we decide to change the location of the cup, we can pick it up and move it through space to the dishwasher; but it will take time to do so.

A CONCEPTION OF THE UNIVERSE

"Now that we have space-time fresh in our minds, let's focus on orbits for a bit and see how they correlate."

    Why do things orbit each other, and what is a gravity well? These two questions are basically one and the same. An orbit is a fall; our planet is constantly falling through space-time into the sun. It would most certainly make it there rather quickly if it were not for the planet's forward momentum. As the earth falls towards the sun, its forward momentum shoots it past its mark, continuing its fall, seemingly forever.

    When I picture orbits, I like to imagine two views at the same time. The first being a "true" view of an object circling a more massive object in space. The second is a nearly identical view, but this one has a "Gravity Well" below the more massive object; like the ball on a rubber sheet experiment. A plane, supporting a massive object, gradually angling down steeper the deeper and closer one observes towards the massive object, until it comes to a point of drastic decline, sweeping under the massive object. A giant funnel cone appearing to drag in anything that falls its way. It is a standard metaphor used to visualize gravity's strength.

    Let us view the dual screen model of a moon orbiting its planet. On the left, we see our natural view in full color, on the right we see a simulated image of two spheres in orbit, floating over a giant funnel cone.

    The moon too has its own funnel, much smaller in this case, yet it does affect the motion of the larger mass; and it could be fun to imagine the small wobble it creates. Nevertheless, for now, we had better just focus our attention on the large center funnel.

    Let us next set our orbits in motion. We will witness the view at a speed of three days per second. Orbits are never perfectly round; we will use a slightly elliptical orbit for our moon. On the left, we see our true color frame: the moon is traveling in its slightly oval, counter clockwise orbit. We can

EXPLORING THE UNIVERSE FROM WITHIN

watch it as it travels around and around, and around, chasing after the earth.

On the right, we see the moon image graze across the surface of the funnel, seemingly falling, with its forward motion carrying it past its mark in an endless rotating freefall. As it circles around the planet, it is difficult to notice much more than the simple circling motion, it appears rather constant and boring.

So, let us substitute our planet with a star and our moon with a comet. We will change the orbit to a wide prograde oval, greatly off centered from the star. As we peer now at the left screen, we see the comet's long orbit. The orbit carries the comet far from the star, only to loop around and swing back in again.

As it gets closer, we can see the comet speed up and make a hard bank around the star, over and over. The closer it is to the star the faster it travels. The farther away the comet gets from its star, the more momentum it loses against gravity's push towards the star.

On our right screen, we see the comet in its orbit as it glides across the funnel's surface. The comet journeys towards its farthest point away from the star. We watch as its thrust weakens in its feverish climb up the outer edge of the funnel, then as it makes its bank back down the slope and regains momentum.

As it begins to get closer to the star, we watch as it rolls back down into the funnel's pull. As it regains its energy, it swings hard and deep into the well, launching itself back out and up the funnel's wall preparing to circle again.

What we watch is the continuous rise and fall of the comet across the funnel plane, faster as it falls towards the sun, and slower as it gradually loses momentum as it rises back up the funnel wall again; up and down, slower then faster, farther away, then closer again, over and over.

The right screen's funnel surface is truly a great way to visualize the strength of gravity; the funnel's slope demonstrates the varying strength of gravity's push, relative to an object's distance to a larger mass. We see the sweeping drop of the funnel representing gravity's increasing force the closer one observes towards a heavy mass object. As the comet travels away from the star, we see it again climb the increasing rise of the funnel surface. It is losing its battle against gravity's push back towards the star. It has no choice but to continue with its forward momentum and swing back down into the depths of the funnel, to regain its lost velocity.

The true color image on the left gives us a beautiful picture of light and motion, it is pretty much what we are used to seeing or imagining. When paired alongside the metaphor on the right, which emphasizes the various strength levels of gravity's inward push relative to the distance from a heavy mass object, with its depiction of the long rising climbs and steep declines of the funnel surface, the two images combine to show the elegant balance between gravity, mass and velocity, created by the orbiting objects while they travel through vast regions of space-time.

"Press your mental 'Save Button'; we'll be using that model in the future I think. But first, remember discussing how an orbit is a fall, and our planet is constantly falling into the sun? And, how it would most certainly make it there rather quickly, if it were not for the planet's forward momentum? Where do you suppose that momentum comes from?"

Now that we have a basic idea of orbital movement, let us ponder this: there are still other forces that effect the rotation of an orbit.

Our moon has been moving away from Earth since the two objects consolidated just over 4.5 billion years ago. It currently travels about 1.5 inches away a year. It really is not much, but where does it get the energy to overcome space-time's push back into Earth's gravity well?

EXPLORING THE UNIVERSE FROM WITHIN

Since we are already in the Lab, let us revisit a previous experiment. Let's conjure up the old "Container, Gel and Sphere Model". We can start at "step three". We have fully expanded our transparent container, super gel and our heavy mass sphere. Our canon ball size sphere has clearly distorted the surrounding gel; we are able to see the focal lensing effect with our lights.

Let's add a smaller sphere some distance away to represent the moon...we should be able to see the gel distortion surrounding our new sphere as well. Next, I want us to give our large sphere a mild rotation of one revolution per 15 seconds. Visualize what begins to happen to the surrounding gel. It slowly starts to drag along, as it follows and pushes with the sphere's rotation.

It begins to churn, circling just behind the sphere's lead, fastest near the equatorial axis. The swirling effect pulls inward at the surrounding gel, warping a vast ring around the sphere. As the current spreads, it expands, with a tidal pull engulfing our moon. Within moments, the wave of energy surpasses the smaller sphere, continuing beyond as the warping becomes less and less defined the further away we look from our large sphere.

Eventually, our moon will be caught up in this circular wave of energy. When the small sphere succumbs to the force, it will begin to drift. Slowly at first, but if we speed up time, we can watch our moon as it picks up speed and falls into its rotation around the large sphere. The longer we watch, the clearer the orbit of our moon appears as it continues to accelerate. Over time, the smaller sphere may gain enough speed to be slowly slung away from the large sphere.

This simple model gives us a dazzling image of the potential energy transfer between Earth and moon. In our model, Earth's rotation feeds the moon the energy it needs to climb the wall of Earth's gravity well and wage against space-time's push towards the massive object it orbits.

A CONCEPTION OF THE UNIVERSE

Neutralizing gravity's effect comes at a price, paid by the loss of momentum in the earth's rotation. The earth's days are getting longer, while the moon is getting farther away; energy expended in Earth's rotation travels through the warping of space-time, casting away its companion.

End of Experiment.

§

***Note:** "Gravity pushes back into mass. If the mass has a rotation, it essentially deflects space-time's energy. If the rotation is strong enough, as the rotational force travels away from mass, its deflection energy not only widens, furthermore it weakens over the vast distances. The two opposing forces, space-time's gravitational push combined with the rotation's deflection, lend to an invisible whirlpool effect; often assisting mass's forward momentum around a larger mass object. What do you think will happen first? The earth will send the moon off into space as it is ejected from the gravity produce by Earth's presence? Will the earth lose all of its rotating spin and come to a stop before our moon is cast away? Do you think the moon could then fall back to Earth? Or, will the tide of the sun's battle between fusion and gravity shift, as the sun expands into a red giant and consumes both spheres? At a loss of approximately 1.5 inches a year, it is just a matter of doing the math."*

CHAPTER FIVE

§

THE CONSTANT SPEED OF LIGHT

OR

LIGHT RINGS AND HUBBLE BUBBLE BITS

"A photon is a packet of energy, a tiny bundle of information. Ones and zeros, code which allows our adapted brains to interpret our surroundings."

Light travels at the very impressive speed of 299,792,458 meters per second, that's 186,000 miles a second or 671,000,000 miles per hour. It's been measured to take roughly 8.21 minutes for a photon to travel from the surface of our sun to the earth, then a tiny fraction of a second more for that light to strike and bounce off our hands then traverse into our eyes.

After it penetrates our eyes, the photon is registered as an electrical signal by our photoreceptors, then the signals are sent to our brains to be decoded and organized. Every bit of light we ever see travels through space and time...it is just a matter of how much, and how long.

The universe is so extreme in size, that meters and miles quickly become microscopic markers of measurement. Light, on the other hand, is very consistent, a "constant" way to measure time and distance. And, let us not forget, the vast majority of light within the universe is ancient.

A CONCEPTION OF THE UNIVERSE

Light travels from a star in a massive growing sphere through space-time. Most of the photons we capture are millions, and more often billions, of years old. Timeless, yet ever so slightly stretched over the many millennia.

Let's move to the Lab. I say we attempt to build a model of light and observe how it travels. First, I ask myself, why do I see light *only* directly? If I look across my backyard, I am aware that photons are bouncing off my fence into my eyes, allowing my brain to generally register the direction, distance and color. However, why do I not see the reflected photons from the fence, which are traveling in all directions other than my own, at that same moment?

Imagine how confusing the world would appear if we perceived light in a form close to how a sound wave travels. It certainly would give our brains much more to decipher.

In our model, we should assume a photon particle to be two-dimensional. A minute point, which is only witnessed directly. So small, that at any side angle it disappears, moving away from us at such speeds that we are unable to capture the photons in our eyes. That is, until they are directly reflected back by something else. Photons travel as individual packets of energy, but they travel in such massive quantities, and are packed so closely together, they appear as well to travel as waves. Perhaps not unlike the surface of a flowing stream, oscillating currents which, when examined closely, are found to be individual molecules made up of even smaller particles.

"Hence the term 'Particle Wave'."

I imagine light as an attack on any given point enacted by photons. A bombardment of photonic energy, either absorbed by the matter it strikes, or reflected away from the point of impact at purely random directions, one at a time, trillions every second.

EXPLORING THE UNIVERSE FROM WITHIN

What if next, we set up our ray gun to emit a light beam one photon thick. It will fire one photon at a time at a rate of ten trillion photons per second and will strike one precise point of matter several meters away.

Let's set up our next container. We will make it into a perceivably infinite size, and make sure absolutely no light is permitted to enter from any outside source. We can lay a flat plane of super steel before us. I have set up our ray gun on a tripod of absolute stability, angled at 45° towards the plane of super steel. By remote, we trigger our bombardment of photons. Each photon strikes one precise point in the middle of our plane consecutively, one photon after the next. We will need to increase our magnification quite a bit through our absolute darkness to where we are able to see a pinpoint of light, a tiny dot lit upon our super steel, imperceptibly waving before us.

Now I want us to view this event through a lens that allows us to slow time, see each photon strike the point, and reflect away, one photon at a time. As we peer at our point, we see what resembles a solid, silver, laser beam. As it strikes the plane at a 45° angle, we can now see each photon slow down and impact our point. Each photon is then reflected in various random directions, one after the other. Each photon looks like a silver needle piercing off of the point, forming to razor sharp tips, blasting out in a growing half ball of silver spines, thickening by the millisecond with a slightly higher concentration of spikes firing directly opposite of our beam (being its brightest reflection, best seen from the opposing angle).

I envision this model as a metallic silver, Long-Spine Urchin, radiating its points in an ever-increasing half sphere of quills, until they direct outward towards every possible location away from the steel. As any one spine is reflected, it immediately begins to fade, and is nearly instantly replaced by another silver photonic spine.

Chapter Five | 39

A CONCEPTION OF THE UNIVERSE

Our nearly infinite number of spines each represent any given line of sight at which one could observe the dot we formed with our beam. All of this assumes we could inexplicably see photons as they travel away from us. This would technically require "light" to be emitted from each photon moving away from us, back towards our direction. That means a photon would have to give off other photons in all directions as it travels.

Since we presume in our model that a photon is essentially a two-dimensional point of energy, seen precisely in one direction only, we know that without our special lens in the Lab, we only see the dot, not our silver urchin of light.

Next, if we were to set alight enough beams to illuminate a one-foot square patch of our super steel, without our lens, we see a bright square of light across the plane. With our lens turned on, we would see trillions upon trillions of spines blasting out in all directions every millisecond, appearing very dynamic and almost life like.

If we measure any given point within the square and count the bombardment rate of the photons, we would notice that there are a certain number of photons that disappear and are not reflected off the plane. These missing photons were actually absorbed into the super steel, and in turn generate various levels of stored energy or, heat, at a rate determined partly by the energy or spectrum level of the photon, along with the impacting matter's elemental structure and surface regularity.

Our model also suggests that ambient light may simply be photons, which are tumbling through space, then into our eyes, after they strike an air borne particle. They may brighten an area when disturbed, but are few and scattered, almost twinkling, simply unable to send us a clear image of their source. We can imagine this spectacle very clearly by picturing the sudden opening of old soiled curtains in front of a sunny window. Large, visible dust particles floating in the air, appearing to glow in the sunshine as photons strike the

EXPLORING THE UNIVERSE FROM WITHIN

large particles and reflect into our eyes, temporarily brightening a darkened room before they settle to the floor.

Let us continue on with our experiment by creating an endless expanse of space-time within our current container. Now bring your focus to an average size star, such as our sun. Let's view our star through a special lens that allows us to follow a band of light only one-trillionth of a second's width, as it's cast away from the star in every direction: essentially a growing sphere of photons.

By separating this small band of light, we will be able to have a clearer view as it spreads across the universe. The first pulse of light we register leaps from its star as a growing, glowing orb. Quickly it grows outward, stretching wider and wider by the millisecond. Before we know it, it engulfs us, then blows past in an ever-widening circumference.

Watch for another pulse. This time as it approaches we will ride along with it for a bit.

"Be ready, here it comes."

Even at our altered view, speeds of approximately one millisecond per second, we are traveling along with our orb of light at an immense rate of speed. In a moment, let's hop off and watch the next pulse from our new perspective here.

As the next pulse is expelled from the star, we have a bit more time to watch it grow, a thin wall of light shaped as a massive growing sphere. Let's stay ahead of it for a while. We will increase our travel rate to send us exactly one light-year, or just under six trillion miles, away from our starting point.

Observe the next series of pulses from our star. We will adjust our perception of our sphere of light to view the orb grow its radius at a rate of one light-year per 5 seconds of our time. We can see the pulses at a rate of one per second jump from the star and spread out in all directions, a slow motion strobe globe.

A CONCEPTION OF THE UNIVERSE

"Best not to stare too long, once I felt as though it may give me a seizure!"

Next, let's fall back a bit further, say, ten light years away from our star. Now it takes about fifty seconds for the light sphere to reach us. As we watch, let's add a few dozen more stars and spread them about at various distances, each pulsing one light burst per second, expanding at a radius of one light-year every five seconds.

Now as the massive, curved walls of light travel, they start to collide and pass through each other. We develop a perspective of how long the light has to travel to get to any particular point in space-time. Let's take a wide circling voyage around our cluster; pick any point within the group of stars and watch as the pulses arrive from the assorted stars and their various distances.

If we observe our cluster from this location, we will witness a bombardment of photons from numerous sources all hitting us at the same time. However, all of the light we perceive at the present moment has actually had to travel great distances to reach us, each pulse leaving its own star at various times, only to reach us, seemingly (from our perspective) simultaneously.

If we change our location to four stars over, we'll notice a different combination rate of bombardment by the various stars. Let's split up and view the light pulses from opposite ends of our star cluster. Ready our watches. Turn off all the stars except the one closest to you, three light years away from your position, and fifteen light-years from mine. We will clearly see the time difference it takes for the wall of light to reach each of us.

EXPLORING THE UNIVERSE FROM WITHIN

There it goes. Within 15 seconds, the light has already reached you and you press your stopwatch. As the wall of light approaches me a few moments later, I will hit my stopwatch as well. We do not need to record the times; we can clearly see that the same moment of light given off from your star reaches each of us at vastly different times.

Next, if we were to spread out and fire up all our stars again, then zoom out so we can observe our constellation in full, we will witness the crisscrossing of all the various spheres of light. From any various point within the cluster, at the very same moment, we would observe a completely different combination of pulses at each varied location.

This is what *relativity* truly means to me; where it literally becomes a matter of when *you* witness an event in time, relative to your location, verses a second observer witnessing the same event, at a different time and in a different location.

"One more point I'd like us to cover quickly."

Our observable universe is just that, observable. It is all of the matter and light we are able to *observe* from our location, it is *not* the whole universe, most likely not even close! The universe has had what certainly appears to be a bit over 13 billion years to allow its light to spread. From any given point in the universe, with an adequate telescope, we would be able to view photons sent from extremely distant galaxies, some of which have been literally traveling for over 13 billion years. Just think about that for a bit!

A CONCEPTION OF THE UNIVERSE

Our greatest range of sight is coined the Hubble Bubble. I view our hubble-bubble as only one bit of data out of possibly trillions amongst the entire universe. I like to think of it as a Hubble Bubble Bit or "HBb." Our bit contains hundreds of billions of galaxies which, due to expansion, have stretched over the last 13 billion years to about 93 billion light-years across.

The size of our HBb is incomprehensibly large. When I consider the possible scale of the entire universe, it sends my mind whirling. Envision our HBb as a microscopic pinpoint amongst the background of the total universe. So incredibly large is the universe, that I imagine our giant HBb as but a drop in an ocean of an ever-expanding infinity.

End of Experiment.

§

Note: "*What if we colored our light pulses? Say, white within the first light-year, then yellow, orange, red, purple, blue, green and so on; a different color for each year the light travels. Now take a position amongst our stars and watch as the changing colors reach you at various locations.*

Also, ponder this: What if you were to stop time and travel through any given point in space? We will imagine you are zigzagging across your back yard. One could imagine, as you turned to look behind you (if you could see at all), you would clearly "see" a distorted view. You did just pass through space as time was standing still. You just carved your path through countless photons, let alone the trillions upon trillions of other air-borne molecules. You have disturbed them

EXPLORING THE UNIVERSE FROM WITHIN

directly, and even formed a wake of increased distortion expanding outward from your movements.

Pushed and compacted, warped and transformed, from an outside perspective, it is a blurring tunnel of frozen, distorted space-time and a cold, darkened, dizzying maze from within. You may just notice that the photons do not travel to you at all, and in order to see, your eyes must go to them, a disorientating confusion of light particles flashing in your eyes.

Now, let's restart time. Poof! Instantly the distortion disappears, not even a blur; light is so incredibly fast. If we were to conduct the same experiment, at a rate of time set at only 1% of normal time, light would then travel at a rate of 2,997,924.58 meters per second. Even slowing down time by 99%, I imagine our vision, although dimmer, would return to a perceivably normal state, as the speed of light would still be extremely fast! Everything around us would appear still, as we are able to move about at our amusement without the potentially nauseating distortions surrounding us.

As useful as disrupting photons may be, within our current model we will not concern ourselves with this effect while we travel. We will simply pass through the photons and matter before us. In our model, our physical presence affects nothing, unless we wish it to.

Oh, and when you're done, don't forget to turn off the lights."

CHAPTER SIX
§

MATTER CANNOT GO FASTER THAN LIGHT

OR

WILL "WE" EVER GO FASTER THAN LIGHT?

"Matter (protons, neutrons and electrons etc.) or energy appears unable to "naturally" pass through space-time faster than the speed of light.
Matter disperses space-time, and it takes massive amounts of energy to disperse it quickly, relative to light-speed."

The farther space-time is pushed away from any given location, the more intense is the force of its wanted return. The curvature of space-time around mass is the condensed force of space-time. At rates of very high density or velocity, mass receives a powerful returning and opposing force.

Pushing an object made of mass at velocities close to light speed uses so much energy, pushing or attempting to tear a void in space-time, that the seemingly imperceptible "drag" or perhaps, friction, of viscosity from space-time eventually builds sufficiently to set the speed limit for mass. It would almost appear that space-time ultimately becomes an equal and opposite contrasting force, barely noticeable, until intense extremes of density and velocity are present.

∞

A CONCEPTION OF THE UNIVERSE

With that said, the voids within the universe are very, very cold. An extremely steady temperature of approximately 2.7° above absolute zero, or 2.7° Kelvin, it is what is left of the cosmic background radiation; a wash of photons left over from the Big Bang event which, since the colossal event, have been steadily cooling over the eons.

Of course, we are able to witness areas of extremely high temperatures across the universe, but oppositely so, there are nebulae that are measured as being as cold as 1° Kelvin. Yet, there is only one location known in the universe to have been colder than 1° Kelvin and, one might say, it was not completely natural to the universe.

We (this planet's current conquering species of animal, Homo sapiens), in laboratories on earth, created the phenomenon. Experts have procured a method of capturing a certain amount of matter within a magnetic field, holding the molecules so tight that their motion nearly stops at microscopic levels. No movement, no energy released, no heat.

I make this point in order to give emphasis to the fact that we are at one possibly incredibly rare location within the universe. Only here, within our planetary system, is it possible to reach such extremes, seemingly unattainable anywhere else in nature.

Thus far, to our knowledge, we are alone in the universe (as utterly ridiculous as that sounds considering the ultimate scale of the universe we inhabit). With enormously vast distances between habitable systems, it must be a rare event indeed for there to be communication between two alien species with more or less the same level of technology; let alone the chances of just one of the species overcoming the impossibilities of safely navigating the stars for a face-to-face encounter.

Therefore, as it stands, we are the lone species capable of obtaining the knowledge, the resources and eventual ability to reach extremes which cannot be met anywhere else in the

EXPLORING THE UNIVERSE FROM WITHIN

known universe. At least by the current nature of the universe...that is to say, by "mother nature" alone.

"That is, if indeed one, mistakenly, does not consider the human species to be a part or force of nature."

What possibilities lay ahead for the advancements of human science and technology? With enough time and success, the possibilities are virtually endless.

"And, that's not a 'metaphor,' that is surely a fact."

Even the ultimate speed limit of space-time could possibly be reachable, opened and explored by our species. It is not unimaginable to consider someone seriously pondering, one day in the future, whether the "constant" itself could perhaps be persuaded to actually be exceeded by mass! Just think of how fast our technology is advancing...it could be sooner than you may imagine!

Sadly though, it is much more likely not to be for hundreds or maybe even thousands of years, if indeed it is at all possible. There are times for me when one hundred years seems far, far too long to wait! Nevertheless, let us take a moment to think of the advancements made by a species that is fortunate enough to have a hundred thousand (or even a million years or more) to study the sciences. They (or *we)* would truly be "god-like" compared to the primitive simian species we are today.

However, for now, the human race must vehemently continue our efforts to survive and move past this precarious moment in our evolution, a tempestuous period in which our species has finally begun to take the true first steps out of the "dark ages." A point in our history just before a mass global enlightenment of education, technology and human rights. A time in which our species may have an opportunity to grow past ancient myths and rooted ignorance; where one day we

A CONCEPTION OF THE UNIVERSE

will *all* truly thrive in the advancements and comforts of our future engineering, sciences and common sense based morality.

Taking into consideration the long history of our proven, and sometimes-astonishing, survival instincts, I have genuine hope our heirs will indeed, over time, evolve into and emerge as, not perfection, not without flaws, but still as a highly enlightened and satisfied species. A society that "masters" its resources, safeguards its domain and achieves a fundamental acceptance for the mutual well-being of all life.

"Or at the very least, all life that has the potential to suffer."

A people that continues to carry the ancient primal urge of exploration, and continues the quest to learn and expand their knowledge. It is extremely likely that our successors will indeed open possibilities we have yet to even imagine.

§

__Note:__ "Despite what the history books may say, we are, clearly, still in the tail end of the dark ages. If our species is able to survive this complex period of trepidation, this age of transformation from 'Fundamental Ignorance' to 'Fundamental Advancement'; our descendants may not only rise up as masters of their solar system, but one day truly claim the entire galaxy open for exploration."

CHAPTER SEVEN
§

WHAT IS A BLACK HOLE?

OR

WHAT IS 100% DENSITY?

*"You won't be sucked 'into', the black hole.
The black hole is already full."*

In our gravity model, we considered the equilibrium of space-time, and how the balance between matter and space-time causes the force of gravity's push. In our next experiment, we will take a closer look at density.

Let's set up a container of super gel. We can expand it to an infinite size. We will be able to monitor this experimentation quite safely from our observation deck within the container. We will need to acquire a grain of our super steel, and we can transport it about 10,000 kilometers, or 6,200 miles, forward from our current position.

We will expand our grain to match the mass of the earth. As it expands, watch the distortion of the gel. At our current distance, we should be able to easily detect the change building in the space-time around us.

From our position here, we can clearly see the density curves in our gel surrounding the earth mass sphere. Let's give our ginormous new sphere a slight rotation...we'll make it one revolution per 24 hours, just like home. Watch a minute more; can you visualize what the spin does to the gel?

Along the equator of the sphere we can see a new twist, the rotation pulls at the gel, dragging it along. A large swirl begins to form, it is a wave of energy reaching out in a mirrored whirlpool effect, largest at the equator and funneling its way to the poles.

Our sphere is big, heavy and dense! Hard as rock, one could say. Weighing in at the equivalent of 6 times 10 to the power of 24 kilograms. On the other hand, maybe we should think of it as *receiving* 6 times 10^{24} kilograms of force from space-time's "push"...it is indeed massive.

For fun, let us add another sphere, approximately the same mass as our moon, and place it approximately 240,000 miles away. We will let it assume a "normal" Moon like orbit around our large sphere; it should flow along quite nicely with the gel.

As huge and massive as our Earth-size sphere may be, just like the earth itself, it is still approximately 99.999996% empty space. Between the protons, neutrons and electrons of each atom exists a vast emptiness. You and I are made of such materials as these; we too consist of mostly space.

Our next step in this experiment is to remove all of that wasted space. Time to bust out our trusty ray gun. This time we will blast the sphere with a powerful shrinking ray (it will actually be more of a condensing ray)...give the trigger a pull and let us watch our sphere begin to miniaturize.

We are condensing our sphere as is, we are *not* taking away any of its mass. Being that our sphere is an equivalent mass to Earth, let us think for a bit about what it really represents.

First, I suggest we pluck out all of the approximated 7.1 billion humans on the planet, which is over 300 million tons of bio-matter! Now let us condense (*painlessly of course!*) all of their elements to the approximate diameter of 0.635 cm...or about a ¼-inch size clear marble.

"We'll need to set our marble aside for a moment, for we have much, much more to harvest."

EXPLORING THE UNIVERSE FROM WITHIN

Next, we need to gather all of the remaining proteins, chlorophyll and attached molecules from the planet. That is all of the earth's mammals, reptiles, fish, birds, worms and bugs, plankton...even bacteria, viruses and so on.

Of course, we will need to gather all the leaves, trees, sea plants, crops, grasses, even graveyards, soils and fossil fuels. Let's just make it quick and grab all organic matter on, in and above the planet. We will sterilize the entire world of all life, then squeeze all of it into our wee little marble.

"If we would now analyze our marble, we would find it has grown to be very, very dense. Yet, believe it or not, it is still mostly empty space."

Afterwards we will transport what is left of the atmosphere: all airborne water vapor, clouds, rain, snow, gases! All of the world's hydrogen, carbon dioxide, oxygen, helium and many other remaining molecules. We will grab all of the water from the pools, ponds, swamps and streams, rivers and lakes...we will need to empty the Great Lakes, along with many massive underground aquifers worldwide. Then on to the oceans! We must also remove all of the snow from the mountaintops and greater latitudes. Let us not forget to free the artic caps of their ice. We will strip the planet bare, and it all goes into the marble.

I suggest we set the marble on our planet's surface for a brief moment. Its mass will alter our planet's rotation and even its orbit, right before, and as, it begins to sink straight through the surface of the crust. Within minutes, it should sink right to the core of the planet. It is extremely heavy, but it still has plenty of room! It is still not even half-full.

"This is what I mean when I say, 'Our atoms are mostly empty space'."

A CONCEPTION OF THE UNIVERSE

With the marble now at our earth's core, we can allow it its freedom. The marble will now begin to absorb the planet itself from the inside out, starting with the inner iron core, then outer core.

Finally, our marble has begun to grow outward, as it swallows the molten lower and upper mantles. Then she consumes the roughly 5 to 30 mile-thick crust of solid earth that encases the planet...the entire supply of the world's gold, platinum, silver, copper and iron ores (even the densest diamonds are crushed beyond recognition). Trillions of tons of heavy metals, all of the other various elements and minerals. Then whatever is left of the cities, all of the world's skyscrapers and industrial buildings, along with billions of homes and various vehicles.

We will also want to collect what remains of the schools, hospitals and stadiums, along with the entire remaining man made infrastructure, including millions of miles of roads and tunnels. After all that, we'll need to consolidate the deserts and mountains, flatlands and ocean beds...all that remains of the 13 billion, billion, billion, pounds of mass.

"That's a 13 with 27 zeros after it."

All of it, concentrated into what started as one single ¼-inch sphere...our now very "Black" marble. We have just crushed the entire planet and everything on it into one of the densest points imaginable. Our marble has now grown to nearly two inches in diameter. It is an amazingly wondrous material, pure matter, 100% matter. Zero percent space, zero percent time, nothing but pure matter, pure energy.

Now is where we can really begin to see the beauty. Just look at the gel surrounding the tiny sphere. Our gorgeous black marble is surrounded by massively distorted gel: distorted space-time. However, here in the Lab, we are able to look beyond space-time's distortions and its pitch-black

EXPLORING THE UNIVERSE FROM WITHIN

shadow, a shadow that shrouds the little sphere in darkness and obscurity.

The mass of our marble is the exact mass of the earth. Therefore, we must assume the push of space-time is exactly the same as it was prior to the earth's recent concentration, only now much more tightly condensed around a much smaller center of mass.

Another brilliant detail to consider is the marble's rotation. Like the classic example of a figure skater leaping into a spin, then pulling in their arms, causing their momentum to surge, our marble too, has "pulled in its arms". One might say, to an infinite degree. In addition, I would say we would be hard pressed to now measure our marble's current rotation.

"But it may be fun to try to calculate though."

Let us view the surrounding gel again. It has become incredibly distorted. We are now witnessing the force of space-time at its fullest. We have created a point of zero space-time, and space-time is not happy at all! It "demands" to return to its equilibrium. It cannot stand the fact that it is unable to squeeze, even the tiniest percent of itself, into this sought after location. Matter has conquered this place. As small as its claim may appear, matter is the dominating force at this location. It is enough density to raise a massive amount of havoc. More than enough to displace an enormous amount of space-time from its coveted symmetry.

Our marble's rotation is simply amazing. It is swirling at a blinding rate, so incredibly fast its movement appears as an indescribable illusion. Just a beautiful black point, a singularity. The pressure, or the "push" from our gel is truly tremendous, matched only by the "current" let loose from the marble's rotation; it churns and swirls space-time's torrential assault.

Chapter Seven | 55

A CONCEPTION OF THE UNIVERSE

It is a massive force...get too close and that will be all she wrote. Before you even recognized your peril, it would be too late. You would be swept away by space-time's undertow, only to coast onward to your doom. It is a line of distance from the sphere called, "The Schwarzschild Radius." It is the definitive point of no return.

It is the radius at which space-time is attacking so fast and viciously, light itself is swept up and cannot escape, even at its brilliant velocity. This would be the *shadow* of darkness surrounding our sphere, giving our marble a much larger appearance to an external observer.

In truth, anything other than light would have been trapped much sooner. It is a whirlpool of tremendous space-time currents, currents that relentlessly attack and steal away any form of energy presented, in order to wear down its nemesis.

Nevertheless, in its rage, space-time is actually feeding the marble, making it stronger. Space-time's push traps and sends any particle within reach straight to the surface of the sphere.

It is a battle of cosmic Jiu Jitsu. A practitioner of Jiu Jitsu will often use their opponent's own strength against them. If one wishes for you to come closer, they may first push you away. As you return your energy, in an attempt to remain balanced and possibly even counter, they will then yank you towards them and, quite often, straight into their trap.

Mass, too, will use its opponent's energy to its own benefit. Take a second to gaze at our marble. Peer past the distortion and its lightless shadow. Focus just on the marble. Imagine it as slowly growing around its edges. Watch the edges soften and blur. Space-time literally smashes the surrounding light, and any other particle, straight onto the surface of our sphere; building it outward, layer by layer.

As space-time attempts to recover its lost location, its push is out of control...it appears furious. Diving full force

EXPLORING THE UNIVERSE FROM WITHIN

into our little black sphere, the push will capture and smash every bit of information possible. Every particle unfortunate enough to get within its grasp...be it photon or atom...every particle is smeared into a Planck-sized film across the surface of the sphere, essentially energizing its foe.

The rotation of our sphere is now so blindingly fast, it appears to welcome space-time's assault, gorging off any particle thrown at it. Then it jettisons the gel (*space-time*) north and south to its poles in the returning circulating current produced by these two mammoth forces, churning relentlessly, all in order to carry on the brawl.

Let us not forget our moon. Surprisingly, it appears unaffected by the battle it is witnessing, since the mass it was orbiting has not changed a bit. Granted, there is a considerable concentration of "current" surrounding the marble, but our moon is far enough out of its reach it need not be bothered. As far as the moon is concerned, nothing has really changed.

So then, let's give it a shove! Let us send this little beauty straight down the well. As it approaches, it begins its new *orbit,* caught in the current of space-time. From our position, we can watch as it circles our black sphere. Slower and slower, until it reaches the Schwarzschild Radius.

At this point, we can only step back and watch, as the moon begins to be pulled apart. Particle by particle, a new stream of energy is being slowly drawn into our sphere. Until, finally, the whole image begins to fade away, to eventually and very slowly disappear into the shadow.

I would suggest we also contemplate the trip as a passenger on the moon. From the moon's surface, our view would be a bit different. We would ride a faster and faster current, orbiting the marble and its looming shadow at an accelerated rate...only to be rushed into an ever-receding shadow, shrinking quickly, as we follow the surrounding photons onto the surface of the sphere, hurled, very rapidly, to our end. Particle by particle, atom by atom, along with the rest

A CONCEPTION OF THE UNIVERSE

of the moon, we will be pulled apart and smeared into a plank size film, across the surface of the marble.

We would not be sucked *into* the marble; the marble is already filled to, perceivably, 100% capacity. When provided with more energy, it can now only grow outward. However, with enough time and a serious lack of added fuel (including light) the marble will begin to lose mass. Mass that is converted into energy and released into space-time, as matter and space-time grapple for position.

End of Experiment.

§

Note: "*In this model, the Black Hole is a state of pure matter, a location that is timeless, bizarre and, seemingly, a glimpse of an imposing and diverged dimension...one which we could think of as, "out of this universe." Yet, if we were to enter, it would not take us anywhere. I imagine as every particle of us is slammed into the sphere's surface, and every bit of information within us is torn apart and flattened, nothing of "us" will remain. Yet, simultaneously, every bit of our data is trapped and contained... seemingly forever. Even the images of our last moments will not escape. I dare say, not even our "souls" would avoid capture. Every particle of matter and energy we consist of will be sealed within for an eternity. Perhaps, also, we should take a minute to imagine if, next, we were to send in a star. How would the star look as it was delivered to the marble's surface? How large would the marble grow with its added mass? Is the spark of fusion extinguished within such densities? What if it is not?!*

EXPLORING THE UNIVERSE FROM WITHIN

Imagine, too, if we sent in another marble!"

CHAPTER EIGHT

§

HOW CAN THE FINITE BE INFINITE?

OR

HOW CAN THERE BE NO CENTER?

*"Inside the universe there is only one true center...
and it's everywhere."*

We ask, how can there be this so-called "Big Bang"? That's like an explosion, isn't it? Yet, I've heard there is no center of the universe. How can this be? The event had to come from somewhere, did it not? No matter how it's shaped...be it a ball, saddle, plane or donut...it had to start somewhere. Right?

From any given point in the universe, every galaxy that is not gravitationally locked to another galaxy appears to be moving away from one another. If you take that information and reverse the motion, it should crunch back down to a center point. Correct?

If everything outside of our galaxy is moving away from us, we must be in the center of the universe...the gods made the earth and then created the universe around us. Right?

A CONCEPTION OF THE UNIVERSE

*"My reply to all of those questions would basically be fairly consistent across the board.
Yes! And no! But, mostly no."*

Yes, if you stood just outside the Milkyway Galaxy and rewound the expansion of space-time, you would eventually see all of the other galaxies racing towards our Milkyway, coming together in a big crunch. However, the view would be the same regardless of which galaxy you chose to observe this event from...and from that perspective, wherever one may be, they are the very center of their own universe.

Once more, *yes,* we are in the center of the universe, but so is everywhere else in the universe!

"I'll open up the Lab."

Again, if we were to reverse the expansion as we just discussed, we would indeed witness all of the matter in the vacuum of space collapse in around us. But, as we stated, the same is true from any point of reference within the universe. From any given location where we would choose to observe the event, the view would be approximately the same.

"Let's begin a short experiment."

Envision being inside a tiny microscopic box in the middle of nowhere...and I do mean nowhere. So small that you are confined to and fill every void within the box. Nothing else exists.

Now, expand the box by a factor of 10 to the power of 20. If *you* were to expand along with the box, you would still take up all the space within the box. From your perspective, nothing would have changed: you are still confined within the same space.

EXPLORING THE UNIVERSE FROM WITHIN

Imagine that the box is really a tiny, tiny ring shaped box and you fill the entire area contained within. If next we expanded the ring to the factor of 10^{20}, but this time you yourself did not expand; where would your perspective of the center be realized? I would assume, at the very location from which you observed or *perceived* the event, since you just saw the interior of the box expand away from you in every direction. Since our perspective would be the same from any point inside the ring, where would the center of the ring actually be located? The real question to ask is, where did the "Big Bang" event occur?

"The answer is... outside of the ring, left behind, at time-zero."

The location of the "event" and center of the universe are two completely different subjects, just as a floating bubble has a center, yet the center was not where the bubble was created.
So if this were the case, and you were within the ring shaped box, from your perspective, there can never be a true center of the ring...only what you *perceive* as the center.
From an observational view outside of the ring shaped box, we may see the center of the ring. But, the center is not within our universe. If we reversed time, we would be able to see the center point where the inflation began. Where the ring would not only shrink, but also travel back to: the original location of the big bang event. Within our model this point or, location, is actually *outside* of our universe; thus unreachable to anyone. Not because of a magic dimensional barrier, but due to the extreme rate of expansion...let alone the possible density or dangerous conditions of the medium in which our universe travels and expands.

A CONCEPTION OF THE UNIVERSE

∞

We should also cover the idea of an infinite universe. I suggest, nothing can truly be infinite, something can only be "perceived" as infinite...you can always add one more. Yet, I would also contest, our universe is indeed infinite. How can this be? It is a paradox which is easy to explain. It is not only a matter of perception, but also actual physical limitations.

The answer? Combine great vastness with a continuous expansion force. The space-time in-between any location and the "outer universe," continues to race away from any observer faster than the speed of light, making it, seemingly, impossible to pass through any barrier or "edge."

Consider if we were fortunate enough to detect the edge of the universe, then subsequently attempted to travel towards it. We would chase after our mark forever, falling farther and farther back as time goes on.

Imagine our current universe model, shaped as a ring. The universe is not within the rings interior circle; the universe is located within the structure of the ring itself. To an outside observer the ring may appear two-dimensional, only as a flat ring. From inside the structure of the ring, an observer would witness 3-dimensional space, plus time.

When we look to the sky with the Hubble Space Telescope and other various instruments, we are able to see light that has traveled for over 13 billion years...nearly since the time the first stars were born. With a bit more information, we are currently able to estimate the age of the Universe to be approximately 13.8 billion years old.

Therefore, if light has had over 13 billion years to travel through space-time, we may assume we could calculate the fact that if we can see almost 13.8 billion light-years one way and 13.8 billion light-years the opposite way, the observable universe must be about 27.6 billion light years

EXPLORING THE UNIVERSE FROM WITHIN

across. And, that *would* be the case, if space-time, too, were forced to constrain its growth to the speed of light.

Matter apparently will not naturally move faster than light...but space-time can! Space-time itself is not restrained by the same laws as matter. It can, and will, stretch and carry the matter and light contained within itself, moving at much higher velocities than that of the speed of light.

This is "Expansion," or, "Dark Energy." Since expansion of space-time carries light and mass as it travels, we look to the heavens and witness views of galaxies whose light was carried over 13 billion light years still currently observable from nearly any direction in space. We are actually witnessing galaxies that are now at distances of over 46.5 billion light years away, moving away still faster and faster by the second; giving us a view of an observable universe that is now known to be stretched by expansion to a diameter of approximately 93 billion light years across. As we look back in time, through space-time, we witness the ancient light of long distant galaxies. Their light continues to this day to leave the very same galaxies at this very moment...light which will never reach our view in the future, due to the rapid speed of expansion.

The space in-between the most distant currently observable galaxies and our home galaxy the Milkyway has expanded to such a size, that future expansion will be at a much greater rate of speed than the speed of light...carrying the light away from us faster and faster, and faster still, than the light within can travel in our direction.

In other words, given enough time, from the perspective of an observer in the far distant future, in the skies where we *now* see galaxies, our observer will see red-shifted views of fainting galaxies...until they eventually fade away into the darkness. Leaving an observer with a view of a very lonely universe: a universe containing only the stars and matter located within their home galaxy; and who will most

A CONCEPTION OF THE UNIVERSE

likely perceive their galaxy as being alone, and the total observable sum of the universe.

End of experiment.

§

Note: *"What if we were both confined within the ring shaped box? Evenly distributed 50/50, your head at my feet, my head at your feet. Can you visualize our different perceptions of the center's location during an expansion event, as we are led apart?"*

CHAPTER NINE

§

WHAT WAS THE BIG BANG?

OR

DID THE BIG BANG REALLY GIVE US SOMETHING FROM NOTHING?

"The 'Big Bang' was not a 'bang' at all. Nevertheless, it was a moment of a tremendous reaction, one which dispersed an infinitely dense point of matter and, by a means of processes, re-created the material into the elements we see today."

Imagine a model of the Big Bang in which matter, matter that formed the initial plasma wave within what we now call space-time, then settled and cooled to become the first protons, neutrons and electrons (predominantly Hydrogen atoms) along with photons and some other various forms of particles/energy. This material did not appear by magic, it was not actually created at the moment of the "Big Bang." Instead let us imagine that it already existed in some unbelievably massive quantity and form within a Multi-verse. Packed perceivably infinitely tight into what we could call a perfect mass of pure matter (a pure, beautiful matter that is infinitely dense), to the extent that any given microscopic point of this condensed material would be considered a singularity, if it were found within the confines of space-time.

Consider if the "Big Bang" was actually the dispersal and expansion of an infinitely dense collection of matter, in a matter vs. anti-matter reaction. One which, after all the "dust settled," not only created the growing void of space-time, but also provided the void with enough energy/matter to form the vast amounts of primordial elements which coalesced to give

A CONCEPTION OF THE UNIVERSE

birth to the first stars and, ultimately, galaxies. Igniting the chain of fusion that is responsible for the heavier elements, along with all of the elements we ourselves are made of and observe in the universe today.

Let us move to the Lab. We need to create a new container. This time it will be shaped like a microscopic light bulb. Within this bulb is another slightly smaller bulb, just the right size to fill the outer bulb nearly completely, leaving between the two bulbs an extremely thin space. A void.

I suggest we fill the void with our Super Gel. Inject it into and around the entire inner void between the two bulbs. Once we have the void completely filled, we move on and completely fill the interior of the inner bulb. Although the inner space of the inner bulb is enormous compared to the super thin void between the bulbs, the space is tiny still. We will need to fill it with a new, beautifully perfect, dark blue, infinitely dense gel from our arsenal. We'll call it "Expansion Gel".

First, imagine handling this infinitely dense material. We will need to set up a workspace. I suggest we create an infinite vacuum completely absent of everything...no matter, no light, no gravity and no space-time. Possibly, from our perspective, infinitely small and large at the same time. Now we should be able to handle our infinitely dense Expansion Gel. When the inner bulb is full, we can seal both bulbs.

Next, to keep it symmetrical and somewhat familiar, I suggest we pull out the old ray gun and give the bulbs a zap. Doing so, we will manipulate the shape of the bulbs to near perfect spheres, one within the other.

We now see our beautifully round and completely transparent (unbreakable at any size) sphere perfectly suspended within the middle of our vacuum. Let us increase the illumination from behind the sphere and shine a light through it towards our direction...it shows us a spectacular light green ring silhouetting a dark orb.

EXPLORING THE UNIVERSE FROM WITHIN

Through the very edges surrounding the dark sphere, we see what appears as an incredibly thin atmosphere of our lightly green Super Gel. It, of course, represents space-time within our model. Pan the light around to the front of the sphere and we can see through the layer of Super Gel into the deep dark beautiful blue Expansion Gel inside the inner sphere.

I think we're set. We will monitor and control the size of the sphere at every stage. Let's start by shrinking our already microscopic sphere to a point that we ourselves are unable to truly visualize or fathom, even within the Lab.

The sphere is *exceptionally* small and dense; even the word *infinite* does not do the orb justice. Next, we need to imagine an, "event," which sets our sphere in motion. With even our best super high-speed cameras, we still fail to capture what happens next. With a, "flash," (instantly, and not just relatively so, but what certainly appears to be absolutely instantly) our sphere grows to an enormous scale, engulfing us and leaving us behind, deep within the center of our Expansion Gel. As the microseconds tick by, we are enveloped deeper and deeper within the Expansion Gel. We have not changed our perspective; we continue to be precisely at *time zero*.

Freeze the expansion, an all stop. We can now look around through what is still, and will be for a very long time, an infinitely dense expanse of our perfect, beautifully blue, "Expansion Gel." Although we paused the expansion, we are still already unimaginably deep within the center of the sphere. From this perspective we will not be able to see anything more than our infinitely dense Expansion Gel. Perhaps at this point we could rename this remarkable material "Dark Energy."

I suggest we now transport ourselves to an area contained within our still extremely thin micro layer of Green Super Gel. Once inside, we can now consider our Super Gel as

Chapter Nine | 69

A CONCEPTION OF THE UNIVERSE

our beloved space-time. A newly born, three-dimensional expanse of blazing plasma.

At that raging moment of expansion, called "inflation," subatomic particles ignite into a monumental fusion event. I have heard it has been measured that the "event" was over within the first three prodigious minutes. It is hot here, hotter than we could ever imagine. Pure energy has been ignited and has created a void between what is now two surfaces of inner and outer collections of anti-matter. The "reaction" which ignited the inflation also generated a force that set the sphere in motion...not only in an unparalleled moment of growth, but also in a direction away from the event. Ever so "slightly," to a range of lesser, but still (by any standard) infinitely dense pressure, within an ever greater expanse of anti-matter.

With our model still paused we can travel through the plasma. We are timeless, witnessing the first photons created by the intense temperatures of fusion. We can continue to travel through the plasma...but I have been here before, there is nothing more here we can see. Everywhere we travel within the plasma void will look the same, and it will be so for some time to come.

When visualizing the reaction field, I envision an "inverse bubble." It may appear as a bubble made of pure fire. Maybe, more like a giant, paper-thin sun, traveling through an absolute expanse of Anti-matter. A bubble made, not of soap, but of pure plasma; created by a matter vs. anti-matter event. I visualize our plasma bubble not suspended in air, but instead within anti-matter. As a soap bubble would contain air, our plasma bubble has encased its own infinitely dense dollop of anti-matter...fuel for long-term expansion.

Everywhere we travel through our field, our super gel, our space-time, was only a moment ago infinitesimally small. From a point so dense, everything we now see within our space-time (and then some!) expanded. From that infinitesimally small point. Everywhere we travel, all came from the same event, the same diminutive point.

EXPLORING THE UNIVERSE FROM WITHIN

Let's get ready to hit our "play button." Moving forward we will now adjust our time settings to a speed relative to 10,000 years per second. Push the play button. After a few minutes, we look out to see the glow of the plasma start to fade, its intense heat starting to cool. As the moments tick bye, the light grows dim; first to a thick haze, then the glow begins to completely fade away in lessening, broken waves of dancing photons.

"Then, Darkness."

From every perspective here in the void the view is pretty much the same. Everywhere, at any given moment. Currently we are able to navigate at super speeds through the cooling reaction field, but still never see anything more than what we do now: blackness. It will be so for a bit to come yet.

For now, all we can do is wait, as we stare into the deep. We are being carried along by space-time, imperceptibly traveling at speeds unbelievably fast. Although we cannot see any visible light, we can still use our technology in the Lab to visualize the energy surrounding us.

If we turn on the lab's "particle viewer," the vast majority of particles we detect seem to be racing away from us. Yet, we can start to see some of the drifting elements swirling in dance across great distances. We witness huge bands of coalescing, subatomic particles begin to build the first structures of the universe. Slowly being pushed together by space-time's gentle gravitational fall. Seemingly never-ending towers of atoms, the vast majority each containing one proton and one electron; Hydrogen. Clumping together and collecting in tremendous bands and swirls, the beginning evolution of disorganized order called *Entropy*.

As we speed through time, we prepare to witness the birth of the first observable light in thousands of years. At last, at some great distance away, a light erupts. A ball of gas

A CONCEPTION OF THE UNIVERSE

particles has collected and grown so large that space-time had to begin to crush its way back in, forcing the atoms to be packed more and more densely.

The force of gravity's crushing push from all directions squeezed the gas ball tighter and tighter, inevitably, to the point of fusion; re-igniting the universe with heat and light. A near infinitely dense ball of searing plasma, a battle between the forces of gravity and nuclear fusion...a solar mass.

We can watch as the blast wave of the star's birth radiates across the expanse, causing a collapsing effect within the interstellar nursery (the gas cloud surrounding the "newborn"), condensing portions of the clouds into thick walls, challenging space-time in its perpetual struggle, and daring it to fight its way back. Carrying out its systematic need for symmetry. Forcing a hurried continuance of the battle between huge collecting masses of matter, and gravity's crushing will to remain in its equilibrium.

End of Experiment.

§

Note: "*Again, press your mental "Save" button, for we will expand much more and build upon this model in the next few experiments. Perhaps first we should step out of the Lab for a bit. It wouldn't hurt to walk outside and take a look at the sky for a few minutes.*"

CHAPTER TEN
§

WHAT COULD OUR UNIVERSE LOOK LIKE?

OR

INVERSE BUBBLE THEORY

"Envision being miles deep below the ocean's surface: intense pressure, millions of tons of water over our heads. We watch the ocean floor until we witness a tiny bubble, collected under the cliff edge of an insignificant rock, crushed to its smallest size possible within these great pressures. As the surrounding current shifts, the bubble is released to the sea."

Recently, I've taken to pondering possible shapes of the universe slightly more often than usual, which has led me to take into consideration a few additional factors; not least of which includes the presented evidence that leans towards the curvature of the universe being flat (something like an approximate value of .02%, close to a perfect 01%).

Yet, I am not quite ready to envision the universe as a plane. After mulling over probably far too many possibilities, I have come to the following as my go to model. Let us briefly step back into the Lab.

Imagine a universe shaped as not *just* a bubble, but an *inverse* bubble, where infinitely dense matter and a relatively lesser (y*et still a very infinitely dense!)* anti-matter collided, and a "reaction" occurred, creating a sphere shaped pocket resembling the liquid that fills the space between the inner and outer surfaces of a soap bubble.

A CONCEPTION OF THE UNIVERSE

Imagine still, that during the reaction between the extremely dense matter and anti-matter, our universe bubble encapsulates a fraction of the surrounding anti-matter, and is thus "filled" with the surrounding, slightly *less* dense (but still infinitely dense) anti-matter, at pressures equivalent to the surrounding pressures at the collision event...all contained within the inner surface plane of the bubble.

This is to say, our universe would not be *in* the bubble, it would be contained *within* or, in-between, the inner and outer surfaces of a sphere shaped pocket. In other words, the universe is not contained *within* the bubble, but rather *is* the bubble itself. Rising towards lesser pressures in an unbelievably vast "sea" of anti-matter.

∞

We too can picture the universe as flat, the same way we can picture the earth as flat. Let's take a stroll, shall we? As we proceed with our leisurely walk across town, apart from the occasional hill or slope, we seem to be traveling on a flat surface. The scope of the earth's surface, relative to us, is such that we cannot differentiate the curvature of our sphere. We simply cannot tell the earth is a sphere from its surface.

"At least not without some serious calculating."

As such, it is possible to visualize the plane of the universe so vast from our perspective that it too appears flat.

"At least not without some further serious calculating."

EXPLORING THE UNIVERSE FROM WITHIN

End of Experiment.

§

Note: *"Think back to our bubble on the ocean's floor. As the tiny bubble is jettisoned upward, it begins its expansion. Upward and outward, growing in size with every meter it rises, the pressures gradually allowing the gas molecules to stretch, as the bubble moves towards lessening densities."*

CHAPTER ELEVEN

§

CONCEPTS OF DARK MATTER

OR

THE CRASHING OF WAVES

"An Arctic Ocean wave travels across the surface of the sea. It rises to meet and collide with another wave returning from the shore, several meters away. As the forces join, they swell into the air and crash down upon a series of raised boulders protruding from the sands.

In the freezing temperatures, the splash and mist of the waves collect as layers of ice over the top of the great stones. Very quickly, the thickening layers of ice build up; encompassing the rock, increasing its appearance. The small boulders 'grow' to much larger scales over the weeks to follow, eventually collecting into what appears to be a single mass.

The boulders' true sizes have not changed, yet tell that to any passerby who decides to walk across them or any ship that happens too close."

Straight to the Lab. Let us briefly take a look at Dark Matter. It is said that all of the visible matter we see only accounts for about 4% of the universe's energy. Approximately 22% is in the form of "Dark Matter," and the remaining 74% is the force called "Dark Energy."

"We'll ponder again on Dark Energy, after this experiment."

A CONCEPTION OF THE UNIVERSE

Dark Matter. I have heard it described as free-floating particles that are attracted to matter; particles which lend support to mass against its battle with expansion. When experts scan the heavens, they are unable to find enough visible mass to justify the continued clustering of matter, as it does over great distances.

The force coined "Dark Matter," is called so because it is believed it may be a form of matter which does not interact with light; subsequently being why we can visually detect it, but not actually "see" it.

Let us take a quick look at a possible explanation of Dark Matter. The force of Dark Matter acts as a scaffold in space-time. It lends support to mass, holding the mass together in its various clumps and veins.

Let us expand a container of Super Gel. We will make this one infinitely large, and within it we will spread matter throughout. Scientists have mapped a large portion of our observable universe; they have made a simulation model that looks like a purple glowing sponge with areas of various densities and voids.

Where the universe was once very uniform in its early structure, it has since congealed into pockets of seemingly empty space, surrounded by gargantuan size regions of random towers, branches and streams of consolidating matter. Surprisingly, all the matter we can detect still doesn't correspond as enough mass to resist space-time's force of expansion. There is simply not enough mass to account for the gravity to support such structures alone.

Let us imagine this web of branching matter within our gel. Reach in and pluck the webs of the universe.

"I picture it as vibrating harp strings."

Science can detect dark matter through its effects of gravitational distortion; I see our vibrating strings as that distortion.

EXPLORING THE UNIVERSE FROM WITHIN

Imagine if expansion causes the cycling of vibrations, creating waves, gravity waves; faint waves that crash upon the already distorted space-time surrounding matter. Imagine our container still expanding: the empty voids of the universe become larger and larger, minute-by-minute, while the structures of mass throughout develop and become ever more consolidated and noticeably defined.

As the gel expands, it continues to seek its equilibrium against matter: in this model, constant adjusting of the space-time units sends waves or, vibrations, across the gel. The waves eventually, and continuously, crash into the walls of matter, creating an extra push into matter from all directions, accumulating over great distances as extra visible distortions in space-time, creating the substructure that distributes added support to visible matter.

End of Experiment.

§

Note: "*What additional ideas can you add about this invisible or 'dark' matter? If it is not gravity waves, what is it really? There are current projects in the works where groups are attempting to view and record true gravitational waves. I very much look forward to ascertaining how their conclusions compare with this model, and adjusting the simulation accordingly.*"

CHAPTER TWELVE

§

CONCEPTS OF DARK ENERGY

OR

WHY NO SPECIES WILL EVER SEE BEYOND THE EDGE OF THE UNIVERSE

"We briefly discussed the captured 'dollop' of anti-matter, our 'Expansion Gel.'
In this model, the fuel of the universe's expansion is that dollop.
As the universe bubble travels to a region of lessor density within the multi-verse, the inner dollop expands, equalizing to the outer pressure surrounding the bubble.
That expansion will continue until the end; it is not unlike our tiny bubble on the ocean floor."

Within our model, expansion, or "dark energy," could possibly be thought of as an inner pressure; a force from within the inner circumference of the bubble. As the bubble moves away from the collision point, the density of the surrounding anti-matter lessens, allowing our bubble to match pressures from outside the outer circumference of the bubble shaped void.

The reaction could be thought of as the initial energy that "gradually" caused the bubble beginning to drift away from the direction of the collision point, permitting the pressure within the bubble to equalize, relative to the surrounding pressures of the outer circumference of the bubble; thus causing the seemingly ever-increasing expansion of the universe.

A CONCEPTION OF THE UNIVERSE

The pressure conversion would likely be at a relative constant, with seemingly slower periods due to the "crampedness" of the early universe, and the effect gravity played within those close quarters during the early expansion period. Our model suggests expansion is undoubtedly the greatest force found within the universe, unfurling the entire sum of mass within the cosmos throughout an ever-expanding plane of space-time. In our inverse bubble model, we could theoretically circle the universe around and around forever. In reality, if we were to try to go inward or outward, even at the speed of light, we would fail. Space-time would always stay ahead of us.

However, if we transported ourselves to the outside of our universe and were able to observe the bubble amidst the *sea* of anti-matter, I suggest we might still see the first photons that were ever let loose, from the reaction of the "Big Bang." I envision the outer and innermost surfaces of our inverse bubble, its "shell," to still glow bright with primal radiation. Photons expelled at time zero, the oldest in the universe, impossibly too luminous and torrential to ever peer beyond with mere human eyes. Alternatively, perchance *oppositely* so, shifted red and fading, as expansion diminishes the concentrations of photons expelled from the event.

End of Experiment

§

Note: I claim this is what "infinity," truly is. An exceedingly ever-growing, finite abyss. Our 'inverse bubble universe'. Yet, has there ever been a bubble formed that won't eventually pop?"

CHAPTER THIRTEEN

§

IMAGINING THE MULTI-VERSE

OR

DO WE EXPAND INTO NOTHINGNESS?

"Remember when we asked, 'What's the multi-verse look like'? We just simply don't know...but let us attempt to imagine it anyway."

If we were to postulate that space-time itself has set the limits for matter within the universe, what is there to stop us from imagining that "outside," of our universe, space-time as we know it doesn't even exist?! The limits set by space-time may *only* occur within the field of the Big Bang effect. This, in fact, gives us enough wiggle room to ponder and picture a "god's eye view" of the Multi-verse. Yet, it is not all that straightforward to quickly achieve a strong mental image or model of possible scenarios leading up to the moments before, during and after the "Big Bang." Nevertheless, it is possible to conceive of, and eventually achieve, a full but tentative, hypothetical model of what the Multi-Universe itself could resemble. Perhaps even what it may be contained within!

A CONCEPTION OF THE UNIVERSE

Let us open up the lab once again and turn up the lights! We're about to witness the greatest possible battle imaginable, between what we would consider as truly infinitely dense "Matter," vs. infinitely dense "Anti-matter"! An event which sparks a reaction amongst two inconceivably dense forms of energy, two forms of pure density (at scales and temperatures yet undreamed of) under physical laws which may allow such densities to be oddly fluid...perhaps not completely unlike a form of liquid metal. Although it surely would not be anything close to what we would consider as a "liquid," within our field of space-time.

"It's somewhat of a mind bending assault on one's sense of reality."

Let us begin by forming a massive ball of infinitely dense matter, the size of which is nearly indescribable (well beyond our "normal" concept of infinitely large). I picture it as an elegant, black egg yolk, collected and submerged, suspended within an extremely dense and beautiful deep blue sea of anti-matter.

Where the two surfaces meet, the threat of a reaction occurs. Akin to a tablet of sodium bicarbonate in pressurized water, it sits and waits for the seal to be cracked. At the moment of the reaction, it comes to life, in an instant, and commences to froth and bubble.

Do your very best to visualize an unimaginably massive and repulsive current or charge between these two forms of matter; one that every so often builds to a breaking point. To such an incredible degree it must release its grip in a momentary lapse, whereby, in the "tiniest" of fractions and for the "briefest" of moments, the two forms of matter "interact."

EXPLORING THE UNIVERSE FROM WITHIN

Infinitely dense matter colliding with slightly less, yet still infinitely dense anti-matter, will do much more than simply "froth and bubble." Imagine a reaction that ignites a "thin," incredibly intense, energy field: a void in or where the reaction took place. A void which we will one day call *Space-Time*.

This "reaction" is brief and tremendously violent, incomparable to anything I could describe. Once triggered, an incredible annihilation begins. An intergalactic "god-almighty fusion"; a paroxysm which consumes presumably all of the activated anti-matter, and nearly all of what we now call matter, into a point of pure energy. All of which is contained within the event's quickly expanding reaction field, igniting our "paper thin" sphere of plasma. All in a blazing instant.

By postulating a slightly less infinite density of anti-matter (which is like comparing a googolplex squared vs. a googolplex cubed or something else as mind boggling), we can somewhat justify in our model the remaining matter left within the universe after the Big Bang event. The "small" amount of matter particles left over after the reaction, which inevitably accounts for all of the visible matter we see today. All within what is now the expanding field of the "reaction" caused by the "Big Bang" event...the territory we now regard as space-time.

In the *End* (obviously after at least 13.8 billion years and, with a bit of luck, long after 13.8 trillion years), regrettably, our model predicts the inevitable demise of our universe, as well as every other universe ever formed in such a fashion. Ultimately, the bubble reaches a nearly "pressure-less" area of the surrounding anti-matter, the outermost edge of a surrounding sphere of anti-matter, where the universe will reach a barrier of negative surface tension, and is then annihilated.

A CONCEPTION OF THE UNIVERSE

"Pop!"

As the universe collapses, the remaining matter is then exposed to the surrounding anti-matter, and another "minor" reaction occurs. The matter contained within is dissolved into its most basic elements of pure energy, within the surrounding anti-matter of the multi-verse.

We could further continue to postulate that the left over "residue" energy generated from this final matter vs. anti-matter collision could, over "time" and future processes, be reconstituted back into its original matter or anti-matter particulates. Conceivably, recombining with their original sources, to start anew.

End of Experiment.

§

Note: "What would happen to any existing Black Holes at the end? Would their encounter with anti-matter spark a series of mini- bubbles, perhaps a boiling, effervescing foam? Hundreds of billions of 'tiny,' short lived universes?"

CHAPTER FOURTEEN

§

WHAT COULD HAVE CREATED THE MULTI-VERSE?

OR

IS THERE MEANING WITHOUT A MAKER?

"Kahunkna, the sun, became so very jealous of Lunisica, his father the moon's beautiful brightness, that once whilst in a rage, Kahunkna cast a mighty star into Lunisica's heart, whose blood then poured out to form the earth.

Tears arose from Lunisica's eyes, as the loving father considered this betrayal brought on by the hand of his son, and he wept, over what he thought of as his poor scion's anguish. The tears fell to the earth below, bringing forth the rise of mankind, along with the various beasts and flora.

Kahunkna watched as his father's brightness faded into the earth. Then witnessed its great beauty once again, as the world ascended around him. Kahunkna became so ashamed of his actions, he hid his face from this new world. In his remorse, Kahunkna promised his fading father he would return every morning to nurture and care for his new mortal brothers and sisters.

A CONCEPTION OF THE UNIVERSE

In the end, Lunisica was pleased by his new creations. Our glory to him provided the strength for the great father to remain with us and among the gods, where he can watch over and protect us. There are still days when we, his children, please him so, that he shares with us his crescent grin."

Just imagine all of the unending possibilities our minds could conjure, ingenious explanations of ideas and ideals; told in the hope of quenching a relentless quest of curiosity, by a species filled with endless questions and infectious wonder. Accounts told of ancient alien encounters, ghostly spirits and super natural powers, alternate time-lines, other dimensions and even parallel universes, along with the most magnificent tales of myths and gods.

Legends and traditions created in an attempt to satisfy life's great quandaries and perhaps give meaning to our most celebrated achievements (as well as our moments of suffering)...to provide purpose and hope to our sometimes-anguishing existence within this amazing world.

We may forever continue to ask questions such as: "Why are we here? What is our purpose? How did it all begin? Why is the world so cruel? What happens to us when we die? What if nearly everything we thought we knew to be true, in our hearts and minds, has just been plain wrong? Worse yet, what if we are never able to bring ourselves to recognize it?" In the end, we must continue to ask ourselves, "What more can we add to our models? How has the evidence changed?" And, most importantly, "is it yet time to rethink them all?"

"All good questions to ponder.
Perhaps, I'll go to the Lab."

EPILOGUE

Ψ

WHAT IS MORE INSPIRING THAN BEING CREATED WITHIN A STAR?

OR

LIFE IS A PAINFUL AND PRECIOUS OPPORTUNITY

"In an otherwise seemingly infinite collection of inanimate primordial elements, life formed, and has risen out from within our Universe...ultimately evolving to wonder enough to gaze back upon its self."

The very same fusion processes which have created all of the elements we witness today (that's all of the matter we can possibly observe in the present universe: the sun, our earth, all the planets and moons in our solar system, every star in the galaxy, all of the contents of each and every one of the billions of galaxies that we are fortunate enough to witness), the very same processes created the very same atoms with which we ourselves are created.

Our bodies, our very molecules, down to every single atom contained within us; all were fused deep within long dead, primordial stars. Created in the course of the lives and deaths of the stars themselves. We are not simply placed here *within* the universe. We are not simply *in* but are *of* the Universe...namely, *of* our earth!

A CONCEPTION OF THE UNIVERSE

"We are not called Earthlings because we were born here, but because we were born 'of' here."

Again, we are "of" the universe. Just as much so as all of the air molecules stirring outside our windows, and the trees swaying in their breeze. The birds nesting within their limbs, Earth's mountains and oceans, our entire planet itself, along with our magnificent, life-giving sun. All of the blazing stars, all of the hundreds of billions of galaxies which we are fortunate enough to observe today, and as much so the trillions of trillions of possible swirling galaxies we have never witnessed and never will.

We are so damned infinitesimally small in the vastness of it all. Yet, amid all of the great quantities of mass we are capable of observing around us, *we,* are amongst the rarest and most precious bits of matter the universe has to offer. Bits of inanimate matter, forged deep within massive energy events.

Matter fortunate enough to not only *just* exist (which is amazing in its own right), but also to *live,* and eventually become capable of, in fact, looking back upon, and marveling at, its very *self.* To, at long last, stare upwards towards the heavens and behold the great vastness of an infinitesimally small portion of the shear infinite awesomeness assembled before us, all contained within our universe... or, we could say, contained within *us.*

Let us read again the first lines of this epilogue. It says:

"In an otherwise seemingly infinite collection of inanimate primordial elements, life formed, and has risen out from within our Universe...ultimately evolving to wonder enough to gaze back upon its self."

EXPLORING THE UNIVERSE FROM WITHIN

This concept (as well as many others involving our lives amongst this universe) when first realized (whether as a child or an adult) begs our very souls for deeper inquiry. How did it all begin? What *is* the meaning of it all? We should also ask ourselves, "Why do so many of us presume we deserve, or are owed, our very own self-righteous destiny?"

What if there is no grand design after all?! If not, then the responsibility would rest solely on *us* alone to create our own meaningful existences and legacies...a notion that can be an overwhelming concept to us in our weaker moments.

I question whether we, as a species, are yet ready to except the exquisite mysteries and breathtaking truths of reality, without the placebo effect caused by the allure of legendary and inspiring, deep-rooted yet fictitious and often irrational, traditions; which have evolved symbiotically throughout our species. Many of us are conditioned to be frightened by the uncertainty of whether or not we could ever be genuinely satisfied with life, while conceding to ourselves all we truly have is this incredibly brief moment in time.

Our one and only chance to witness and attempt to fathom the awesome magnitude, along with the wonders and horrors, of a seemingly all-magnificent cosmic epic. A "classic," an amazingly charming and powerful, action packed comedic tragedy. This immensely scary, painful and tremendously glorious fleeting opportunity, this miraculous "adventure" we perceive as existence.

End of Experiment.

§

POSTSCRIPT
Ψ

"Daddy, why is the moon following us?"

When we travel under the skies, the moon does indeed appear to follow us. It is so very far away from our eyes, that large objects such as trees, buildings and especially clouds, inevitably become momentary obstacles...which from time to time obstruct our line of sight. I, too, still ponder that grand illusion even today. I receive so much enjoyment when I imagine all of the wondering eyes which have stared skyward toward our enchanting moon.

Our families and friends, both near and far, all of the dear loved ones we have lost over the years, all of the souls we've treasured, each and every one we've ever loved or admired within our lives and throughout history...our heroes and our champions... together with the billions of those souls here on the planet with us now whom we unfortunately will never have the opportunity to meet. A myriad of experiences we can scarcely even begin to imagine.

Those, too, across the far distant past; our ancestors, all interconnected over thousands of millennia. Every creature on this planet who was ever capable of raising its face towards the heavens; all life, that has ever risen out of the earth. And, yet still to come, all of the future generations which are to be born under its watchful gaze...our beautiful moon has, and will, follow them all.

∞

"Just as it follows you."

OBSERVATIONS

OBSERVATIONS

OBSERVATIONS

A CONCEPTION OF THE UNIVERSE
or
EXPLORING THE UNIVERSE FROM WITHIN

© 2015 J.C.V.
All rights reserved.

www.ingramcontent.com/pod-product-compliance
Lightning Source LLC
Chambersburg PA
CBHW030813180526
45163CB00003B/1269